I AM WITH YOU

The Earth Speaks to Us

by Sten Linnander

To Fred and Barbara,
May we continue to be
connected so deeply
through time and space.
Love,
Sten

ISBN 978-3-9814889-1-3

Published 2011 by Sten Linnander, Enkheimer Str. 15, 60385 Frankfurt am Main, Germany.

Table of Contents

Acknowledgements

I would like to thank the following people for their generous help in making this book a reality:

My wife, Karin, who gave me invaluable help and support in the initial phase of writing this book by insisting on calling a spade a spade and not settling for vague formulations.

Virginia (Gigi) Coyle, for her empathic clarity of mind and heart, and for her feedback, allowing me to see the little things that matter.

Julio Lambing, who enthusiastically supported my endeavor and whose untiring intellectual rigor and historical and cultural insights helped me avoid several pitfalls and encouraged me to make several important additions.

Stephan Steinhäuser, for his support during the early stages of this book, guiding me with his professional skills, knowledge and feedback.

Stephen Davis, who provided enormous help with proofreading and designing and implementing the complicated process of turning my manuscript into a published book.

To you all I express my deep gratitude.

Introduction

For many people, the idea that the Earth is alive, sentient and conscious is nothing new; it is simply a fact. Others regard it as a fairy tale.

Many cultures prior to ours held the belief that the Earth is in some sense alive. Ask any Native American medicine man or aborigine chief, and many of them will claim the Earth is not only alive, but consciously interacting with humans. They say this not only because it is part of the belief system in which they were raised, but because of experiences they have had personally.

If we knew the Earth upon which we are walking is a living, sentient being, we would tread on it with a different attitude. If we also knew the Earth can "see" us, that it is "with us" and we can communicate with this being, we would establish a strong sense of care and respect toward the Earth. This kind of awareness and communicative action would have immediate healing effects on the Earth and the human being.

In the 1970s James Lovelock introduced the "Gaia Theory," which – although it does not go so far as to state that the Earth is a living, conscious being – shows that in many regards the Earth is a self-regulating system, exhibiting behavior similar to those of a living system. But he stopped short of regarding it as having any kind of consciousness: *"I am not thinking in an animistic way, of a planet with sentience."*[1] However, today a great many people are doing just that. The modern world view seems to be moving slowly toward the possibility that the Earth might "in a sense" be regarded as truly alive.

One of the few people who have taken seriously the idea of the Earth being not only sentient and conscious, but also able and willing to communicate with humans, is the author John Lamb Lash. He regards the assertion that the Earth is sentient not as a matter to be accepted on belief or rejected because of its unscientific nature; instead, he sees it as a proposition to be tested. He goes on to say, *"Our view of Gaia will stall out in blind speculation unless we can allow that she can communicate with us in language as we know it. Unless this is possible, we will never be able to confirm that she is sentient in the same way animals are, and we ourselves are."*[2]

This is precisely what this book is about. It is an account of the personal experiences I have had communicating with the Earth from within. It describes how these communications came about and includes transcripts of a series of "conversations" I had with the Earth back in 1996, interspersed with more recent material, as well as my own comments and thoughts about what I received.

Taking the communications I received at face value, the Earth is saying that it is alive and desiring to contact us all consciously, both individually and collectively. It goes so far as to propose building a device with which the Earth can communicate with us in our own language. It is also saying that we humans are the determiners of any communication between us and the Earth, i.e. that it will only

---

[1] James Lovelock, *Gaia: The Practical Science of Planetary Medicine*, Gaia Books, 1991, p. 31.

[2] John Lamb Lash, *Not in His Image: Gnostic Vision, Sacred Ecology, and the Future of Belief*, Chelsea Green Publishing, 2006, p. 335.

occur if we are open to this possibility and actively seek this contact. This book, then, is like a ring signal from the Earth, waiting for us to "pick up the phone."

PART I

How It All Came About

Chapter 1

Who Am I, the Author?

I feel that anyone reading a book about communicating with the Earth has a right to first know a bit about the person who engaged in these communications.

But when we describe who we are, we rarely manage to get across more than a very one-sided view, leaving out large core areas that govern our lives. Also, what we have chosen to "do" with our lives is often (but not always) a poor reflection of who we truly "are." An added structural difficulty stems from the fact that we do not know who we are.

So I have chosen to first provide a brief chronology of my life, followed by a description of my quest through life, seen in retrospect.

A Brief Biography

I was born in Sweden in 1950. My father was a judge who spent many years in Africa and Asia as a legal advisor to various foreign governments, and my mother was a senior secretary and teacher who later worked in Switzerland and Africa in the fight against female genital mutilation.

I spent my first six years in Sweden. Then our family moved to Kabul, Afghanistan, for two and a half years, where I began my education in an English-speaking international school. After another year in Sweden, we traveled to Kathmandu, Nepal, where we stayed for three and a half years. There I first attended a Jesuit school for Nepalese children and then an American school.

Back in Sweden I went on to study physics and geophysics, graduating from the University of Stockholm in 1975.

After a year and a half of work, I left Sweden and joined a radical community in Austria. I was to stay there for three years.

Upon leaving the community, I spent the next three years mostly in the United States, acquiring patents for and marketing new German inventions for cleaning oily waste water and other ecologically-oriented technologies.

After that I took a leading role in the development of a social experiment in Germany, working mainly with various aspects of consciousness research. I was the head of a "free university," organizing visits by internationally acknowledged researchers and innovators from the social, scientific, and spiritual areas. I stayed with this project for eleven years.

Toward the end of my involvement with this project, in 1989 I co-founded and led a German non-profit organization, which organized aid programs and technological transfer, and carried out a number of projects in the areas of ecology, medicine, and city administration in the former Soviet Union. We also facilitated an international project for the protection of the Arctic, which was sponsored by the Association of Cities of Northwest Russia.

In 1992 I began studying the spiritual-shamanic teachings of the Sweet Medicine Sundance Path, an eclectic synthesis of the knowledge of medicine men and women of the Americas and Australia that also incorporates modern scientific and psychological insights. I continued these studies until 2010.

In 1994 I moved to Arizona to focus more deeply on the Path, and during the next eight years I also worked extensively with "Energy Synthesis," a theory of fluid self-organization in nature developed by the late Alfred Wakeman. I also established a non-profit research organization and initiated a project to prepare for the worldwide distribution of live high-resolution images of the entire Earth, taken from a satellite that has yet to be launched.

Since 2002 I have been living and working in Frankfurt, Germany, continuing my work with the live images of the Earth and developing and running further experiments relating to Energy Synthesis. To make a living, I have for the last fifteen years been working as a translator from German to English for companies and individuals.

I am married to my beloved wife Karin, whom I first met in 1983; we became a couple in 1984 and were married in 1988. She has been the greatest critic of my work and has at the same time helped me bring it "down to earth" as much as possible.

My Quest

Having grown up both in Asia and Sweden, I was exposed to various cultures, traditions and belief systems early in my life. I realized that the Western culture I was born into, with its beliefs and world view, was one of many, and that what I was brought up to believe was not written in stone; also, that our Western world view, which seems so rational and true, is just a snapshot of what is currently thought to be a correct interpretation of who we are and how the world around us functions. What is considered true in one generation is replaced by entirely different and often opposite "truths" in the next one.

Take the physical world around us. The "objective" theories of the world I encountered in my studies of physics and science in general did not satisfy my wish to understand the world we live in, for the world itself seemed not to have anything to do with me and my life, nor those of others. It was a world in which we are all surrounded by dead matter we can manipulate, but which will remain forever dead.

According to this view, we are far and wide the only living, conscious beings – except for animals, which are simply primitive versions of ourselves, so we cannot expect any intelligent answers from them. We seem to be encapsulated in a world that is indifferent to us, and only other humans share our fate. We are forever out of touch with the rest of the Universe, at least until any ET's arrive or until we manage to find them somewhere, but that will probably take a few more centuries. Even then we can't be sure they won't simply kill us and take all our money.

My studies of physics dealt with the nature of the material world around us, the nature of matter – of which all our bodies are made – and ultimately the nature of the atom. So what was the innermost insight about what matter is that science had found out, even proved? It was that matter is made of atoms, which are made of protons, neutrons and electrons, which are made of quarks and leptons held together by bosons…. Yes, yes, but what are they? The kinds of answers I received were always an abstraction, such as: They are "probability functions." Pardon? To me, they might just as well have told me that matter consists of "maybe valleys." It didn't resonate with anything within me at all. Ultimately, I was told to just do my math like everybody else.

As I switched in midstream from studying theoretical physics to geophysics, I heaved a great sigh of relief. Now I was dealing with the Earth, with clouds and lightning, earthquakes and tsunamis, rock formations and minerals, rivers and oceans, jungles, deserts and swamps, plate tectonics and the inner movement of molten lava within the bowels of the Earth. Although what I learned was purely scientific, the Earth came alive to me.

My quest continued. So who are we? Who am I? Some said I was an individual human being. Period. I was born, I lived, and then I died. Again, period. What about humanity? I was a part of humanity, and humanity somehow existed as a being, a species that evolved but did not really have a consciousness of its own. It was more like a school of fish that seemed to act like a living being, but in reality was just the result of the biological programming that evolution had hardwired into each individual. So when we all do the same thing, all of us together seem to act as one being. Although we aren't. The unfathomable enormity of the Universe around me and the incredible tininess of viruses, atoms, protons, quarks and gluons were just the final icing on the cake, compounding my ignorance.

Then there was the whole spiritual overlay. Again, who are we? Do we have souls? Do we reincarnate? Are there Gods or Goddesses?

Are the mountains alive with spirits? Is a river holy? Was Mohammed the prophet of God? How was I to know?

Then there was the Christian religion that seemed at least to be based on love, forgiveness, hope and redemption. Nice as those qualities were, in history the followers of Christianity seemed not to have read those parts.

Then there was my father, an avowed atheist, whose attitude toward Christianity could be summed up in his statement, "Have you heard what they say: 'In the beginning was the Word?' I've never heard anything so stupid."

I think most people, especially the young, at some point go through a phase when they seriously try to understand who they are and what's up in the world. But these questions remained with me – some would probably say a bit longer than was healthy.

By now other things were bothering me. As the years passed and I grew up, I increasingly realized my own inner feelings and well-being were largely determined by my own emotional structures, and I felt I had little control over them. This was not nice at all, especially since I was pretty depressed and did not know what to do with my life. Sure, I could look around me and see that others were not really so different from me, but that was scant comfort. At least they seemed to have accepted or didn't care less about the explanations about who we are and the meaning and purpose of it all, and were prepared to simply find their place in society. For some reason, that did not work for me.

So what to do? I knew I had to do something pretty radical to clear up the combination of restlessness and depression that kept me like a revved-up engine with the brakes applied. Changing that was a prerequisite for anything else I wanted to do.

Having completed my studies and worked for a year or two, I decided to join a radical community in Austria. It was not a New Age community, but rather a place which threw out the old "nuclear family" values in search of a more alive, actionist and artistic lifestyle – with common property, "free sexuality" and spontaneous, expressive theater. After three, for me tumultuous, years, which included going through the community's own brand of Reichian therapy and intense energetic emotional work to clear out my system, I left feeling liberated and free of my chronic inner turmoil.

I also came away with the feeling that the attempt to create societal structures that emphasized the importance of community was important, and that any viable future will need to find ways to satisfy the basic human needs that throughout history were fulfilled by tribes, clans and extended families and today are sorely unfulfilled in Western society and increasingly everywhere on Earth. On the other hand, collective structures often risk sacrificing individual autonomy and free thinking, and this was certainly the case in the community I was a part of. Overall, my stay there definitely had its light and dark sides.

It is all too easy to leave out the real reasons for what motivates us and sets us out on new paths in life, so I should add that two of the most positive upheavals in my life occurred when I fell in love. The first time was when I was in the middle of my, for me uninspiring, physics studies. It came out of left field and within five minutes – at 1 a.m. on May 3, 1973 – I fell in love, and my life changed. The next day I decided to quit my studies, began to write poetry and make short movies, drank gin and generally walked around on clouds.

The second time was at the Austrian commune, where I had an "analyst" whom I really liked a lot. At one point they decided it was a good idea for the "patient" to fall in love with his/her analyst and use that as a motivation for their "healing." That was an easy task for me, and it cracked my armor; again I walked the clouds. These peak experiences did not last for more than a few months, but they set me off on new directions in life.

Free to pursue my interests, in addition to arranging for patents and marketing new ecological inventions in the U.S., I spent several years traveling, visiting psychics and healers, people sensitive to earthquakes, and other researchers involved in unorthodox studies of matter and the living world. I studied the work of Nikola Tesla and Viktor Schauberger and began what was to become a thirty-year collaboration with the late Alfred Wakeman, a little known artist and industrial designer who developed a fluid dynamic theory of the Universe he called Energy Synthesis.

During my many years with the German social experiment, I focused on various aspects of consciousness research. Having more or less by chance stumbled across my ability to regress people in hypnotic trances to "remember past lives," I traveled widely to determine if the descriptions of these past lifetimes were historically accurate. In several cases the detailed information I had written down from these sessions proved to be factual far beyond what anyone could have invented or checked out in advance. To me this did not represent definitive proof we have lived before, for there were several possible alternative explanations as to how this information could have been accessed.

Ultimately, I came to believe that yes, we probably have lived before. However, I believe that the "we" that experiences this is not the normal everyday "we," but a part of ourselves we rarely include in our daily definition of who we are, and whose memory we only access under extraordinary circumstances, for example after a great deal of practice or through wisely designed ceremonies.

Based on my experience, I am also convinced a great many reports of "previous lifetimes" are simply inventions of the subconscious mind.

Having facilitated over one hundred such "reincarnation trances," including many within the German project, a number of fascinating interrelations came to light. If taken at face value, they indicate that we often spend time with the same people in different lifetimes, as if together we have taken on certain issues to work on through several lives and that groups of people somehow "decide" to come together at different times in history to continue their work together.

I had read extensively about the subject, but unearthing these stories firsthand gave them added poignancy. In one instance a man and a woman provided detailed complementary versions of a relationship they had in Carthage over two thousand years ago. They did this without having knowledge of what the other person had said, since I conducted the trances back to back without the presence of the other person. For me personally, these experiences served to expand my view of who we are, but were not necessary or even productive when it came to dealing with issues in everyday life.

This expansion of my world view continued in other areas. One experience involved sending a ghost on its way, who, it seemed, had decided to hang out in the cellar of the house where we lived. For several months a two-year-old in the community had claimed there was a man in a black cloak in the cellar who "beat the horses." I took the child on my arm and went down to the cellar, and as I approached the corner where the child said the "man" was standing, something hit my chest and fell to the floor. It turned out to be a glass button, and I have no idea where it came from. I was taken aback, sent the child upstairs and confronted the ghost, ultimately telling him he was dead, and that I could help him continue on his evolutionary path. I spoke aloud in German and received hair-raising, perfectly timed answers in the form of loud knocking sounds in the plumbing pipes in the cellar. Having just read accounts of similar work,[3] I called on beings who could help him navigate the transition to other realms and I fused their energies with his and sent them off. The knocking sounds stopped. After that the child told me the "man" was gone and never came back. A few months later we heard that the cellar was the oldest part of the building, having been built several centuries ago, and it had been used as stables for horses.

But my real interest lay more in the idea of a living world from which I felt we are cut off. Within the framework of the Free University at the community project, I sought out and invited the most interesting people I could find to come and speak about their work in lectures and workshops, and sometimes share their insights through hands-on experimentation.

One such visitor was Peter Caddy, the founder of the Scottish spiritual community of Findhorn, which became known because their contact with plant "devas" or spirits had guided them to grow vegetables in the sands of Scotland so exceptionally large that the BBC decided to do a six-program series about them.

Then there was Cleve Backster, originally a lie detector expert working for the CIA, who discovered that plants react to human thoughts and emotions. His work, although disputed by some, also showed that cells taken from humans and kept alive at great distances from the person reacted synchronously with the person's other cells.

---

[3] John G. Fuller, *The Ghost of Flight 401*, Berkley, 1983.

Also on the visitors' list was Jim Nollman, whose continuing work with interspecies communication[4] has made great strides toward understanding how dolphins, orcas and whales communicate with each other, and how we in turn can communicate and commune with them using, among other things, music as the medium. This kind of participatory research helps to reintegrate science into life itself, making it relevant to all of us.

Rupert Sheldrake was a guest as well, and his theory of morphogenetic fields and research on the non-physical connection between pets and their keepers also expanded my view that we are living in a world that is alive and communicating, with a network of communication, "waiting" for us to take our rightful place.

Overall, I pursued developments that led me toward a world view based on wholeness instead of separation, process instead of static existence, interaction instead of action/reaction, emergence instead of causation, and self-organization as opposed to hierarchy and control. Here, mainstream science had made great strides. The quantum physicist David Bohm's concept of the implicate order, Ilya Prigogine's insights into self-organizing systems, and even quantum physics' unusual claim that any observation affects the system being observed, regardless of the specific method used for observation, all opened my mind to a new way of seeing reality.

Taken together, for me these ideas and insights illuminated the many ways that, in terms of consciousness and communication, the "dead" world around us is not dead at all. Nor is it dead to what we do, feel and think. What seem to be dead is we ourselves, dead to our senses and to the impulses that our bodies, minds and souls – as well as those of others and the entire living world – are constantly sending to us.

---

[4] See www.interspecies.com.

My search for answers was always accompanied by new experiences. I attended "spoon bending parties" in Washington D.C., arranged by a U.S. Army colonel, where I saw four-year-olds gleefully bending long, strong steel spoons and forks like spaghetti. The shock of seeing this impossible feat made me suspend my disbelief, and when I gently pressed on the spoon I had spent half an hour trying to infuse with energy, it suddenly felt like putty in my hands and curled up.

I invited Dr. Willard Fuller, [5] an American spiritual healer specializing on healing teeth, to come to Switzerland, and during public sessions he allowed me to stand directly behind him as he did his work. Right before my eyes, I saw amalgam fillings slowly turn into gold fillings.

But I also saw how powerful *disbelief* is, and how fear of the unknown can directly affect reality. A friend of mine and I were playing with a Ouija board in an apartment close to Monte Carlo and were thinking of gambling at the casino. When we asked about this, we were told to "find money." Since we had borrowed the apartment, we couldn't very well ransack the place for money, so I asked facetiously if I could find the money in my wallet. We got a clear "yes." Taking the answer at face value, I got my wallet where we knew I had exactly 300 French Francs that we had exchanged the same day at the border. Together with my friend, I checked the money, counting first a one hundred Franc note, then a second one. The third one, however, was a 500 Franc note. I thought, "This is not possible." I looked up into the eyes of my friend and realized he had seen exactly the same thing. My mind started to reel, and as I looked back down at the money in my hand, I saw the number 500 slowly fade into being 100. I knew at that moment it was fear

[5] See www.willardfuller.com/fills.html.

that had made us revert back to a reality that did not allow for such miracles.

Having read and studied a wide range of literature concerning spiritual matters, parapsychology, etc. – the quality of which was mixed, to say the least – I had the desire to find out for myself what to believe, as far as I could, and at least tentatively what premises to follow in my life. I discovered that this was a fine line to walk. On the one hand, I found incredibly unsound generalizations, exaggerations, unexamined rumor cultivation, etc. and a preying on people's gullibility, most of which seemed to be rooted in a desire for financial gain, status, or simply to be regarded as an "acclaimed" something or other. Correspondingly, I encountered an equally incredible desire to "buy into" such claims without any kind of reality check, and these two groups seemed to have found each other.

On the other side were the doubting Thomases who derived their sense of identity from a refusal to examine their beliefs, especially their "core" beliefs, and to instead "debunk" anything that might threaten the status quo of their world view. Mainstream society seemed to belong to this second group, backed up by powerful interests that also had to do with financial gain and power.

Yet the overwhelming reaction to phenomena that did not fit the existing belief system was to ignore them. "Sound skepticism" was hard to come by. The original Greek meaning of "skeptikos" was an "inquirer" or "examiner" – someone who was unsatisfied and still looking for truth. This requires a balanced mixture of sobriety, openness to the unknown and willingness to question our basic assumptions, including assumptions about ourselves.

All in all, I believe we are now approaching breakthrough expansions of our scientific paradigm to include the dimensions of sentience, consciousness, thoughts and emotions as being as real as the material world, ultimately recognizing that they are different expressions of the same "stuff," and that they interact with each other intimately.

Then there was the Earth itself. Ever since I was a child growing up in different countries, I had a global outlook on life, and I always had a feeling for where I was on the globe. When I was asked to facilitate a project for the protection of the Arctic, I realized the importance and fragility of this area of the Earth. I also began to feel the Earth is more alive than we tend to give it credit for, and reacts not so much to what we do to it physically as to the intent behind our actions, similar to Backster's conclusion from his work with plants and to how we ourselves react.

Later, my many years on the Sweet Medicine Sundance Path opened me up to firsthand experiences of how we can interact from within with the energies and beings of the living world. The teachings and ceremonies of this Path are designed to awaken internal knowledge through direct experience, and they went a long way to reconnect me with a world where not only animals and plants, but stones, landscapes, planets and stars come alive. I had innumerable experiences that showed me that our normal take on reality reflects but a small part of the spectrum of the living world surrounding us all the time, and that we are all so much more than we are taught to believe.

Yet in spite of this, nothing had indicated I would be communicating on a more or less daily basis with the entire planet. I am well aware that vast numbers of people experience conscious contact with our living "Mother Earth" every day, and have been doing so for millennia. My experiences are thus not that unusual. Yet on the one hand I have promised to publish what I received during my communications with the Earth, and on the other hand

I feel a need to share my experiences simply out of a feeling of excitement and awe at what to me is real and amazing. I also believe it is only a matter of time before the consciousness and sentience of the Earth become fully accepted by mainstream society, contributing to a transformation of our understanding not only of the Earth and the Universe, but of who we ourselves are as human beings.

Chapter 2

About My Communication with the Earth

Going Up the Mountain

*I*t was a hot Wednesday afternoon on September 18, 1996. I was sweating profusely as I trudged up toward one of the many summits of a small mountain range within the city of Phoenix, Arizona. As I got closer to the summit I had chosen, I noticed there were other people there, and one of them waved to me. Since I needed to be alone, I decided to choose a different place and veered off to the left. I saw a large bird circling around one of the other little summits and decided to go there. As I reached the top and sat down, looking out over the city of Phoenix in front of me, I was filled with doubts about what I was doing.

I had made my way up the mountain to fulfill a request I had received a week earlier in a trance-like state during one of many ceremonies I had taken part in within the framework of the spiritual-shamanic training I had been pursuing since 1992. The intent of this very elaborately laid-out ceremony was to "remember" a "past life" of mine, and this would be my third ceremony of this kind.

As I mentioned, I had many years earlier extensively researched statements from a number of hypnosis-induced "reincarnation trances" I had facilitated. I came away from this research intrigued by the historical accuracy of the statements made in some of the trances, which often included obscure details requiring much research to corroborate. I do not believe we can ever "prove" we have lived before, but I had decided to keep an open mind and was willing to explore the possibility myself.

During the ceremony, which included a kind of guided trance, I found myself in a place that made me seriously doubt – and I still do – that I was experiencing images of a past life. I knew I was on some kind of spaceship. "This is not a past life and spaceships aren't my thing," I thought; but since it was all I had to go on, I decided to explore where I was. I found myself in a corridor and entered into one of the side rooms. The room was very dimly lit, and some light came from a window at the far end of the room. I walked up to the window and as I looked out, I saw a majestic sight: The Earth, in all its colorful splendor, set against the darkness of the void.

It was a fascinating sight; and as I was taking it in, I heard a voice speak inside my head, "Now that we have made contact with one another through this very elaborate ceremony, the line of communication between us has been anchored cellularly in your body." My experience was that it was the Earth itself speaking to me.

It went on, "I want us to continue our communication. I want you to go up on a mountain here in the area and talk to me, outside of any ceremonial setting." That was all. I was quite astounded at all this, for during the previous such ceremonies I had only been a passive observer trying to glean information about what might have been past lives.

It took a week before I decided to follow up on what I had experienced during the ceremony, and I strongly doubted anything would come of it. But there I sat, finally, up on the mountain, sweating, with a pad of paper and a pen in hand, feeling slightly stupid.

Remembering to keep an open mind, I closed my eyes and returned in my mind to the beautiful image of the whole Earth I had seen in the ceremony. Suddenly, there it was again, the voice. "This is a power spot – you were led here," were the first words I picked up. A few moments later, "I will teach you about the relationship between you (people) and me. This is a program that will go on for two-and-a-half months. Then it will be in place. This is part of our agreement and I am glad it is happening."

Thus began a series of communications that were to change my way of relating to the Earth and our lives on Earth. At one time during the course of these sessions, I pledged to publish my story about all this, and through this book I am finally fulfilling that pledge. It took me a while to decide to follow up on the original request to go up on a mountain and attempt to speak to the Earth, and it has taken me very much longer to finally publish this story.

What Took Me So Long to Publish This?

There were several reasons why it took me fifteen years to decide to publish the results of my communications with the Earth. But the bottom line is that I was not ready to do so until now.

At the time when I received them, I had no doubt I was truly in contact with the Earth. What came through was so obviously real to me, and the statements often surprised me to the point where I felt that I could not have come up with them myself. But during the times in between, I rarely felt the same strong connection. The times on the mountain were quite special; otherwise my everyday life went on without much thought about the Earth.

In spite of my experiences, it has taken me quite a while to truly accept the idea that the Earth is a conscious, sentient and communicating being as reality and not just as a nice option or possibility. Since even I was hesitant about accepting my experiences at face value, I also doubted that I would be taken seriously if I published my story.

I sensed how far-reaching the consequences would be if humanity as a whole would discover that the Earth is alive, conscious and communicating, and if we entered into a continuous, conscious exchange with the Earth. It would have the potential to change our lives at a very fundamental level. I knew I had an enormous fish on my hook; the question was if I was willing to reel it in by fully trusting my experiences.

As I began to write this book, I realized I could not stand behind these communications without now establishing the kind of contact with the Earth I had in 1996. If I could not achieve this, I felt I could not speak convincingly about the Earth wanting to communicate with us consciously. Although since then I had once in a while experienced communicating with the Earth, these communications were usually sporadic and short-lived.

I also wanted to go back and "check" what I had received by asking the Earth more in-depth questions about what I had written down earlier. In addition, the communications from the Earth were phrased in a language that emphasized interactions and processes – including inner ones – over "hard" facts and scientific explanations. It seemed to follow a different inner logic, less of a cause-and-effect logic than the logic of a Universe that is alive, intermingling and co-creating, where emotions are seen as "felt energy motions" and even time becomes "events" that touch each other. I felt that this new way of relating needed a more in-depth explanation.

I therefore decided to rent a small cottage out in nature for six months and dedicate myself to reestablishing as firm a connection to the Earth as I could and write this book. This worked exceedingly well and I integrated the answers I received to my questions into the material from 1996. I now have no hesitations about telling my story.

Some of the statements I received from the Earth are quite straightforward – for example, about it being alive and desiring to communicate with humans, "why humans came to Earth," or a proposal by the Earth that I build a technical device with which the Earth can communicate with us in our own language. Other parts sound more poetic, using words like "the power of beauty" and "the creative power of trust."

Over time, though, I have realized that the "unscientific" parts, once they are thought through to their conclusion, are no less solid than the "straightforward" statements. They simply deal with inner matters for which these words are well chosen. Yet I did feel they needed some "translation," or at least comments. Having shared some of the material with others, I realized many parts were fairly clear to me, who had received them, but not at all clear to others. So I decided to include some of my own comments, thoughts and reactions to what I had received to better convey the meaning of it as I understood it.

For Whom is This Book Intended?

When the Earth asked me to publish these communications, I felt they were intended to be addressed to the general public, not specifically to those spiritually inclined, for example, or those with "New Age" backgrounds and belief systems. If the Earth is alive, sentient and conscious, then it is irrelevant if you believe in God or if you are an atheist, or a pantheist, or simply don't know what to believe. Alive is alive; sentient is sentient; and conscious is conscious.

I realize that for an average person the idea that the Earth is alive "like you and me" is out there at the very fringe of social discourse. But so were the ideas not so long ago that the Earth is round and not flat, that matter is made of atoms, that we have billions of miniscule viruses and bacteria inside our own bodies determining our health, or that invisible radio waves zip through our bodies constantly.

Yet many of the statements I "heard" stretched even my credulity; but then perhaps the best path forward is discovered only when we *do* stretch our credulity. At least that seems to be what is required of us in these changing times.

The Validity of Non-Physical Communication

I wish to add a few comments about how I see the validity of what I received. First of all, I cannot prove that these communications came from the Earth; I can only relate my own experiences. Even if they *did* come from the Earth, I cannot guarantee that I received them all correctly.

I have had extensive experience with various forms of trance work – such as telepathy, remote viewing and various other forms of trances – accessing information mentally or spiritually. Even assuming such accessing of information is real and valid, I am convinced that no such communication is ever 100% accurate. When we communicate with other beings of any kind "from within," the communication is filtered through our subconscious minds. We often do not allow information to come through that we at some level and for some reason do not wish to hear.

For example, a committed peace activist will not easily receive information about the need to take up arms, and a gun rights activist will probably not "hear" statements about the need to lay down his guns. Those who support a creationist or intelligent design view of the Universe will be unlikely to "hear" statements about humans being descendants of apes, and those subscribing to evolutionary theories will have a hard time accepting statements about any kind of divine influence on the emergence of life. Subconsciously we delete, add or distort information so that it better fits with our beliefs, worldview, wishes, hopes, etc. No doubt the same is true for me.

The fact that these communications came through me also means they, to a certain extent, are colored by my vocabulary, my concepts and the categories I think in. These concepts and categories come with assumptions and beliefs about reality that may have limited relevance to the way an entity like the Earth perceives the world, and this may explain some of the more unusual formulations from the Earth that sometimes seem to require a different kind of logic and thinking than what I am used to. In time, as the idea of communicating with the Earth in our language gains credence and becomes more commonplace, I believe these limitations will be overcome.

This book is intended as inspiration, and I hope it can nudge the reader toward becoming more open to the possibility that the Earth is alive and conscious. In my view, the whole Universe is more alive than we give it credit for. If we can relax and allow the thought that the Earth is addressing us, saying, in essence, "I am with you," then this can trigger a shift in us. We can begin to listen to our inner senses that let us feel the constant presence of the Earth – an Earth with which, or with whom, we are closely linked – an Earth that desires nothing more than to be seen, felt, heard and communicated with.

This Is Not a "Channeled" Book

I have experienced many sessions with channels and once invited a well respected American channel for a week to the German community I was part of in the 1980s. I have great respect for channels and their work. But there is quite a difference between the kind of communication I experienced and channeling. Channeling is communication with and receiving information from intelligences in other dimensions, whereas I was communicating with what I felt was a physical living being, the Earth. This is more like communicating telepathically with another human being, with a dolphin or a whale. When channeling, what the channeled entities – be it a deceased person, an "angel," God, Goddess, or any number of non-physical beings – all have in common is that they are not physical. They therefore do not relate to the material world the way we do. In addition, I am not in a trance when communicating with the Earth, as most channelers are when they access their information.

Back in 1996, when my first communications with the Earth occurred, I went back in my mind to when I saw the Earth during my initial ceremony, and then I addressed the Earth. I experienced the Earth answering as a separate, individual entity, and I tried to be as receptive as I could. I only trusted what came to me when I regarded the Earth as a totally separate entity, speaking to me as from A to B.

As I was working on this book in 2011, the Earth recommended I now instead use a different method whereby I connect with the Earth, hold a certain question or desire for communication, and then allow the thinking process within me that wants to happen to occur. I thus establish a state of deep connectedness with the Earth, put out my intent or question, and simply think normally. The feeling I get is that the Earth is thinking within me.

With time I noticed that this process of communication was accompanied by a bodily sensation whereby my pelvic region relaxed and opened up. I took this to be a chance occurrence until I recently read in a book by Dolores La Chapelle[6] that the Chinese regard the "Tantien," the pelvic area, as the "other" brain that can feel the flows in the Earth.

---

[6] Dolores La Chapelle, *Sacred Land, Sacred Sex: Rapture of the Deep: Concerning Deep Ecology and Celebrating Life*, Kivakí Press, 1992.

Notes Regarding Editing and My Comments

The statements I received from the Earth are written in italics and centered on the page.

I have edited the transcripts from my "conversations" with the Earth in 1996, since the sequence of issues that were dealt with were sometimes jumbled. Mostly, though, they are in the same sequence as I received them. Already during my first session, I received an overview of the material to be communicated in the form of headings that I have used as the titles of Chapters 4 through 8.

In 1996 I sometimes asked specific questions to which I received specific answers, but I did not write down the questions I asked. The transcripts therefore mostly read like a monologue by the Earth. I do, however, include the clarifying questions I asked in 2011. To indicate where the answers to these questions and my comments to these answers end, and where the earlier transcript follows, I have added three asterisks (* * *).

Throughout the text I have used the term "we," often followed by generalizations about what humans tend to think about a certain issue. I realize many of my readers will feel these generalizations do not apply to them, and I therefore ask them not to take these statements personally but to see them as my take on what a majority of people tend to think or believe.

Whenever the word "You" with a capital Y is used, it refers to human beings as a whole, to humanity or to the reader. I have used the word "you" in lower case when the word refers to me personally.

Personal note

Since some native peoples in their rituals sometimes use psychoactive drugs to access "the Gaian mind," I wish to emphasize that I did not use any drugs or other mind-altering substances during any of the experiences related in this book.

Disclaimer

Neuroscientists claim that a special kind of injury to the left temporal lobe can make the left brain become disoriented, and the brain might interpret activity within the right hemisphere as coming from another "self," such as a demon, angel, extraterrestrial, or even God.

So if it should turn out that I have told myself what I experienced when talking to the Earth, then I offer this book as a token of my imagination about what the Earth might say if it were alive and if it could speak to us. But I should add that, as far as I know, I do not have any injury to my left temporal lobe.

PART II

The Earth Speaks

Chapter 3

Preface

As I began to write this book in 2011, I connected with the Earth and received the following preface to what follows, first addressed to me...

You are writing this book
for me to be able to speak to human beings.

...and then addressed to the reader....

I have been waiting for millennia
for this moment[7] to come,
when I can speak freely to You,[8]
with whom I am so deeply connected.

---

7 As I understood this, "this moment" refers to this time in history, not this very moment.

8 "You" here refers to humans, both individually and collectively.

I have been nurturing You as my children,
and I will continue to nurture You,
as I have always done.

But now I wish to speak to You
as to a partner.

Throughout the ages we have
spoken to each other in many languages,
but mainly in terms of feelings
and energy movements.

When I speak through Sten and his being,
what I say is always colored
by his way of receiving me.

I wish to speak to You, both individually
and collectively, as a part of my own evolution,
which is deeply connected to Yours.

You have chosen to come to Earth,
both collectively and individually.

And I have chosen to come to You.

My coming to You is an expression of,
and occurred through our joint dreaming.

The transformation that is coming
is a transformation not only of You,
but also of me.

As Your old seers would have said,
"This moment is written in the stars,"
which simply means
that even as we communicate with each other,
we are embedded in an even greater whole
than You and me.

The oneness of the Universe
will always be in a dynamic balance
between individuality and union.

You and me, as separate beings,
coming together as one.

Many of You sense in some way
that I am alive and that I am with You.

But You are not yet allowing Yourselves to
know this in Your body.

There is nothing wrong with this;
it is rather similar to the shy approaching
that is often a part of the beginning
of every adult love relationship.

Chapter 4

Overview, History and Outlook

The Earth – Alive and Conscious

I, the Earth, am alive and conscious.

I act consciously.

I also interact consciously with humans, both with the conscious part of humans and the subconscious part.

My motives are not known to you, because You block out 90% of Your own motives.

As I understand this, it means most of the time we are not aware of our true motives, the hidden agendas that lie behind our actions. 90% might seem to be a lot; but I have often experienced people in existential situations who look back at their lives, describing their careers, love lives, etc., as being driven by the desire to gain the respect and love of their fathers or mothers, the longing to fit in and be accepted, or any number of emotional agendas they were seldom aware of.

If we were to carry a video camera on our shoulder all day long and review it the next day, we would be able to predict with a high degree of accuracy almost our every move, thought and emotion. We would also see the motives behind our choices and that we seldom use our free will, instead being driven by deeply seated programs and habits.

According to the Earth, our conscious thoughts and motives are thus just a small part of what drives us to do what we do and think what we think – the tip of our own iceberg.

I have free will.

Every atom does.

The whole Universe does.

In a sense I am a God – Gaia.

But I am not only Grandmother,
I am also Grandfather.

Most cultures and traditions that regard the Earth as a living being consider the Earth to be female, using words like "Mother Earth," or sometimes "Grandmother Earth." This makes sense, since the Earth supports and nourishes all life on Earth as a mother nourishes her children. I decided to ask about the unusual statement that the Earth is also a “Grandfather”....

Your concept of God and Goddess
is very limited.

You tend to regard God and Goddess as a
kind of super-father and super-mother,
i.e. super-parents; but this is not quite correct.

What You tend to forget is
that what You call Gods and Goddesses
are also inside of You.

This externalization of God and Goddess
has made You either refute the idea
of the existence of God/dess entirely,
in a misinformed search for freedom,
or accept the idea of a purely external
God/dess, or Gods/Goddesses,
to whom You have abdicated powers
that are rightly Yours.

This has become a root of fear,
which can be expressed as a “fear of God.”

When I say that I am not only
Grandmother Earth,
I am also Grandfather Earth,
it means that I encompass more of both
genders than You tend to allow Yourselves.

In relationship to You,
of course I am mainly "mother"
and You have been and are my "children."

I have the same love for You
as a mother does for her children.

I am the womb within which
the seeds of life are planted.

But realize that my womb
includes the atmosphere
and the various force fields surrounding me.

I also carry You and feed You.

I offer to You everything You need to grow
and evolve.

At the core of all physical and non-physical
existence is the union of male and female
forces or energy flows.

As is the case for all of You,
in addition to having a dominant gender,
I also have opposite gender aspects in me.

For example,
the polar areas are more male than female,
whereas the equatorial areas
have more of a female quality.

Yet it is all one.

I am one being,
and I am definitely female dominant.

* * *

I have been conscious and active for eons,
and so have You.

I attracted you,
just as much as You sought me out.

The Earth thus states that we as a species – the collective soul of humanity – were present and active elsewhere in the Universe "before" we appeared or incarnated on Earth.

But are we not children of the Earth? Have we not evolved from the beginnings of life on Earth? If we accept that we have souls, then when a woman bears a physical child, the soul of the child comes in "from the Universe" and she gives birth to it, nurtures it until it is grown, and then sets it free. From this point of view, a child is thus both a child of its mother and may certainly – as a soul – come from elsewhere in the Universe, no matter what this "elsewhere" may look like.

You may wonder if my sentience and
consciousness is a function of yours,
i.e. of human beings' sentience and
consciousness.

But this is not so.

Before you came to Earth
my sentience was different,
but no less real, no less conscious.

Seeing the Earth

Imagine me, the Earth,
as the blue planet in space.

I have not seen myself that way,
but I am that planet.

I have not seen myself from the outside;
that is one of the reasons I attracted you.

Through You,
I can see myself and my beauty,
and I need humans to let me do that.

With humans I have that agreement.

So perception is an important aspect
of the agreement we have with each other –
both You perceiving me and I perceiving You.

As I was going through this text, I asked how it can be that human beings, as the collective soul of humankind, make agreements – and with a planet, on top of it all? I received the following answer....

Human beings make agreements
among themselves.

Wolves in a pack make agreements
about how to relate to each other,
but they also make agreements with other
species with which they live in symbiosis.

These agreements all vary,
but they have one thing in common:
their ultimate purpose is to ensure
an ecological balance.

In this sense, I, too,
have an agreement with humans.

Agreements are an expression
of cooperation toward a larger goal
that encompasses all parties to an agreement.

In the larger sense of things,
they guarantee or ensure respect
for autonomy and individuality
as the basis for cooperation.

This interplay between
individual autonomy and symbiotic oneness
is the growing edge of evolution.

Throughout these communications, the Earth emphasizes that the processes of separation and union – individuation and merging –

lie at the heart of Creation, being central to the development of consciousness as well as to the existence of the material world.

When You see Yourselves through my eyes,
You will understand more about Yourselves,
and of course about me.

Through the space age
we are fulfilling one part of our agreement,
and this is at the same time
the beginning of a new era in our relationship.

Cataclysms on Earth

The following was received in 1996, i.e. before the devastating tsunami of 2004, the earthquake in Haiti in 2010, and the earthquake with subsequent tsunami and nuclear accident in Japan in 2011. What follows came as an answer to a question regarding cataclysms caused by earthquakes, volcanoes, etc. The cataclysms I was thinking of were those that threaten the survival of mankind, such as we see in disaster or "end-of-the-world" Hollywood movies where, for example, enormous waves crash over New York, Tokyo, London, etc., killing off most, if not all of humanity. I was not thinking of regularly occurring earthquakes, volcanic eruptions, and such that have occurred periodically with varying intensity throughout history.

There are some things I will not do,
in terms of interfering with

Your destiny (= Your highest choice),
even if I could.

I will not let my waters wash up
over Your shores in enormous cataclysms,
and I will not erupt,
spewing lava over large areas.

When that is seen, for example in visions of the apocalypse, it is a vision of what would happen if we did not have an agreement – if I did not love You – or if You had colonized me, which I would not have allowed.

You sometimes feel as if You had colonized me, due to Your current destruction of the Earth, but that only goes skin deep.

Although the human being has drastically changed the surface of the Earth from a geological point of view, the impact of these changes on the overall physical system of the Earth is quite limited. The Earth also has an incredible power of self-healing. Should homo sapiens become extinct, it would take only a few centuries for almost all physical traces of human history to be erased.

I have never willfully destroyed any being.

If I have,
it was the being using me as its instrument.

I had a very hard time with this passage. I can believe the Earth does not willfully destroy any living beings. But the statement that those who die in earthquakes, mudslides, tornadoes, floods and hurricanes "use the Earth as its instruments" is a much tougher proposition, and it almost stopped me from writing this book. A few years ago, I watched a report on television about a four-year-old boy who had been caught in a mudslide in the Philippines. He was up to his shoulders in mud, but he was still alive and conscious, and the villagers were scrambling to try to tie a rope around him to pull him out. At that moment the hillside started to move again, and the boy was swallowed alive by the mud. I simply cannot imagine that he "used the Earth as his instrument."

There are many spiritually-oriented people and paths that claim we "create our own reality," and this covers everything that happens to us. I do not exclude the possibility that there may be a state of consciousness where this can be seen to be true, but it is certainly difficult to believe for most of us. Mostly we hear of this statement in terms of "New Age" generalizations or wishful thinking, often supported by references to quantum physics experiments that however seem to apply only to that size range.

If we are ever to "know" and understand that we do create, or co-create, our own reality, we need paths of thinking, experiences and insights that may some day lead us there. Ultimately, they need to be so clear and self-evident that we do not doubt that we all create our own reality, even when we look into the eyes of a dying child, overwhelmed by fear and pain. For me the best case that has been made for the idea that we create our own reality is in the novel "*2150 A.D.*" by Thea Alexander.[9]

As I was going through this text in 2011, I decided to ask specifically if we "create our own reality"....

---

[9] Thea Alexander, *2150 A.D.*, Warner Books, 1976.

Although this might feel very far-fetched
for You in many instances,
the degree to which you know this to be true
determines the abilities that you can access
when it comes to creating Your reality
consciously.

Yet "knowing" is not simply believing.

Knowledge is felt; it is not thought.

Accessing this level of knowledge
is one of your long-term tasks.

* * *

Someone in tune with me
won't be harmed by even the worst volcano
or earthquake.

They would not even be hurt
if the lava swirled around their feet.

I left this rather improbable-sounding statement about not being hurt if lava swirled around one's feet for later communications.

When we see the calamities on Earth resulting from earthquakes, tsunamis, tornados, volcanoes, hurricanes, storms, floods, etc., we often feel helpless. It is as if we are subjected to these forces of nature totally independently of our own thoughts or actions. Is the Earth really saying this is not true?

Human beings have created large-scale systems
that are not aligned with me and my processes.

In Your everyday lives,
You usually do not allow Your energies
to flow freely through Your minds and bodies.

When You start to feel my presence,
You begin to allow me into You, and You
thus automatically connect with my flows.

This makes us flow
in harmony with each other;
we start acting as one organism.

In a sense
You will become an extension of my flows;
You will let me flow through You.

I am an autonomous, individual being,
as are You,
both individually and collectively.

What You call the forces of nature
are an expression of my processes
and those of the living world, including Yours.

So the way to get out
of the feeling of helplessness
is not to try to control or change what I do,
nor simply to try to predict it.

It is to open up
and establish a deep relationship with me
and the living world around You.

This will change not only the way You act,
but the way You interact with me
and Your surroundings.

The more You are aligned with me,
the more we act as one,
in harmony with each other.

So I am not asking You to change Your
thoughts or actions.

I am asking You
to enter into a living relationship with me.

Why Humans Came to Earth

When You came to Earth,
I was no longer a "young planet."

I had gone through many stages of development,
and many of them were expressions
of my own doing, my own searching;
and the same is true for You.

Let us put our existence on Earth in perspective. It is estimated that the Earth is about 4.5 billion years old. The first signs of biological life appeared some 3.5 billion years ago, and humans, or Homo sapiens, originated in Africa about 200,000 years ago. This means that the Earth is about 20,000 times older than the time period during which humans have walked the Earth.

As I write this, I am sixty-one years old. That is just a bit more than 20,000 days. So if I were the Earth, I would only have experienced humans living on me for a single day of my entire life. But somehow it seems it would have been quite an important day in my life.

The Earth is a highly dynamic system that has undergone major changes in its lifetime and continues to do so. As we live our daily lives, we usually do not think of the fact that below our feet enormous masses of hot lava, bathed in intense white light, are constantly flowing in recirculating patterns, slowly recycling the Earth's crust back to become lava, only to be brought to the surface again. At the same time, life as we know it evolved on the surface of the Earth. I asked the Earth about its development before the human being appeared.

My connection with the Sun,
my lovemaking with the Sun,
began a process of conscious awakening in me.

The material expression of this awakening
was the first forms of life on Earth.

Life is an expression
of consciousness materialized.

You were on a quest
to discover new parts of Yourself
and to play in new gardens.

* * *

We chose each other for one specific reason,
which could be formulated as follows…

We were closely aligned,
and we had one common goal:
Allowing, or combining, freedom and intimacy.

We are not used to hearing about common goals in terms of human emotions. But emotions are a part of our reality that affects our whole sense of well-being. We are so used to thinking of the Earth only in terms of its physical aspects that we do not usually "allow" it to have anything corresponding to our emotions.

But have we not felt the "emotions" or "energy motions" of the Earth? There is no denying the emotions that are awakened in us when we experience firsthand the raw power of a lightning strike or a thunderstorm, a raging river or the eruption of red hot lava from an active volcano – or, for that matter, a soothing Mediterranean breeze, or the stillness and pristine quality of daybreak at the banks of the river Nile, or Mississippi, or Ganges.

Yet we usually do not believe that what gives rise to such strong emotions in us corresponds to some form of emotions the Earth itself might have and with which we resonate. I asked the Earth about this....

Human emotions are part of reality;
they are real.

Not just "felt real,"
for what is felt is an energy motion.

These energy motions
affect the reality around them.
I, too, have energy motions that I "feel,"
although in a different way
than how You feel Your emotions.

The Earth is saying that emotions – or "felt energy motions" – are not limited to human beings, and that they play a large role for the well-being not only of humans, but also of the Earth. But how can the emotions of people, individuals, who are so minuscule compared to the enormous size of the Earth, have any effect on the planet itself? Is this not simply an extreme form of anthropomorphism, attributing human characteristics to the Earth, seeing ourselves as the measure of all things?

Not if we listen to those with firsthand experience – to the Elders, the "Grandmothers" and "Grandfathers," the medicine men and the shamans, the native tribes of the Earth, be they in the jungles of South America or the plains of North America, in the Arctic or in the heart of Africa or Australia, or on the islands of Southeast Asia. Not if we listen to the Buddhists and the deep ecologists. Not if we listen to those who listen, to those who feel, to those who actively connect with the living world around us. Not if we listen ourselves.

But what could intimacy mean to the Earth and how can the Earth's intimacy be dependent on its relationship with humans? I decided to ask the Earth for clarification....

Intimacy begins with intimacy with oneself.

One popular description of the word intimacy
is "Into-me-I-see."
This is related to the importance of me,
the Earth, seeing myself through You.

By perceiving myself through Your eyes,
I gain a greater recognition of my own beauty,
thereby gaining intimacy with myself.

When human beings become intimate with
themselves, they open up their conscious mind
to their own subconscious mind.

If I were the Earth – a conscious, feeling, and in some way "emotional" being – then it would represent a great shift for me if I were in close and intimate conscious contact with humans, potentially the most highly developed beings on Earth.

For us humans, the shift would be enormous. Not perhaps at first, for it would take some time for us to stretch our inner attention to begin to feel that we are not alone, but that the planet itself is "with us." We have been searching for life in the Universe for millennia, in modern times hoping for some signal of intelligence from light years away. Perhaps we have been looking in the wrong direction. Maybe we are already surrounded by the conscious presence of the planet Earth itself, and maybe it wants to talk to us.

Our inner separation from the Earth – and from nature – has made us strangers in our own home and made us retreat to the artificial environments we have created. Now we relate mostly to each other and to those artificial and technological worlds. It has also estranged us from ourselves. What I hear the Earth saying is that the more we consciously allow the intimacy and presence of the Earth into our lives, the more of ourselves we will encompass and feel, because we open up to the aliveness of both our inner and outer worlds.

Also, the more we feel the intimate presence of the energies of the Earth, the more we open up to each other. Empathy is not selective. Empathy is one of the greatest peace-makers of all. By cutting off our inner intimate connection to the Earth, we have also cut off our heart connection to other humans and to our own hearts, allowing only a trickle of hope and comfort to sustain us. This trickle is usually not enough for us as a collective to decide to do something about the hunger, wars, poverty and suffering worldwide that we have the means to end if only there was a collective will to do so. It will take a while before we allow our hungry hearts to be touched again. It will take us a while to come home.

Will this contact come about as a living reality for most people? I believe it will, sooner rather than later. I believe it is a part of human evolution, and that humanity is growing up, for the first time facing the planet as a whole.

* * *

Gaining Freedom

I also wondered how the Earth could gain freedom through us....

Freedom is expressed at many different levels,
actually at all levels.

Life itself is an expression of freedom.

All growth is also an expression of freedom.

When a seed breaks through its husk,
this process is an expression of freedom.

Freedom **from** *and freedom* **to**
are different aspects of the same freedom.

For me freedom means entering into a state
where I am integrated into a higher state of
planetary existence, resonating with other
more highly evolved planets.

And humans are helping me do that.

Processes of evolution resonate with each other.

As a child grows up and leaves its home
to become fully independent,
growth and development occur
in the mother and father, too.

This is a joint process.

* * *

Some of Your peoples have experienced
freedom with me, others intimacy;
but the combination has been elusive.

Our Western culture has experienced a high level of freedom in its interactions with the Earth, a freedom which is now global. We have extracted mountains of coal, metal ores and other materials, and we have pumped unimaginable quantities of oil and gas out of the Earth. We have straightened and dammed rivers; connected oceans through canals; cut down forests and replaced them with mega-farms and cities; crisscrossed the continents with roads and railways and the skies with airplanes, missiles and drones; and generally "tamed" the Earth to serve us.

We have done all this without "asking for permission" and without considering that the Earth might "feel" what we are doing to it.

Furthermore, we have accelerated our exploitation of the Earth to the point where the effect we are having on the Earth, whether it is alive or not, is causing changes in our environment that are threatening not only the lifestyle we cherish so much, but the very existence of our own sensitive species. We have "subdued" the Earth as a way of attaining freedom. But this was achieved at the cost of losing all intimacy with the Earth, thus closing ourselves off from a deep part of ourselves. Moreover, it was achieved at the cost of threatening our very existence, for closed systems are not sustainable in the long term and therefore cannot survive.

Yet we come from a long history. Many of our ancestors have communed intimately with "Mother Earth" or "Grandmother Earth," and many people still do. Throughout the ages, farmers, sailors, mountaineers and many who have roamed the Earth have felt an intimate connection with nature, with the Earth, with the oceans and the winds. Entire religions were based on the Earth being alive and in close communion with humans. Throughout most of history, the belief that the Earth is alive, conscious and in communication with us was the norm. The Earth has been revered for much longer than it has been seen as dead.

The beautiful connection and symbiosis that indigenous peoples have with their land and the Earth can make us wonder why we ever left that paradise. The innocent beauty, humility and genuine friendliness of the Nepalese people I experienced in the early sixties as a child growing up in Kathmandu, where they were living in harmony with the land, contrasted strongly with the unsmiling, stressful and complicated life in the West where I came from. Yet slowly but relentlessly our Western ways are taking over even there.

This intimate connection to the Earth has a childlike innocence to it. It seems that we, both as individuals and as a culture, must make the journey of discovery away from the safe and innocent joy of naturalness to encompass even more of ourselves, and from there to eventually return to another kind of innocence and a more encompassing naturalness – this time not as children of the Earth but as conscious co-creators of our reality, caring for and cooperating with the Earth as her co-determining partner.

Today, reconnecting with the Earth cannot mean simply returning to the religions and world views of our ancestors. When humans have connected with the Earth in the past, they have often done so as children connecting with their mother. When we as a species declared the Earth to be dead, it was like rebellious youths rejecting their parents by declaring them to be "dead" to them. Now it is time for us to reconnect with the Earth as an equal.

We need a modern approach; yet any such approach must incorporate the awareness of our deep cellular connection with the Earth, an awareness from which we as humanity have collectively cut ourselves off. It is only when we take our own power, realizing we are able to stand as partners of the Earth in freedom, that we can have a mature and sober relationship with the Earth, a relationship of intimacy and freedom. We will then have gone through a collective, planetary rite of passage.

This has to do with You "seeing me"
consciously and my "being" You.

When we begin to experience the Earth as a living, conscious, feeling being, we will automatically connect with it and resonate with it. Many of us have connected with a loved one so deeply and so empathically that we saw the world through their eyes, in a sense becoming them. The Earth seems to be saying that such an empathic connection to the Earth can make it possible for the Earth to see what we see and feel what we feel.

Or does it already? The Earth is telling us that it communicates with us all the time on a subconscious level. So what difference will it make if this communication becomes conscious on our part? Is a conscious, empathic connection with the Earth on the part of the human being a prerequisite for the Earth to be able to see what we see and feel what we feel, to the point of the Earth "becoming us"?

No, it is not a prerequisite;
it is a part of the process itself.

I can only see myself when You "see" me.

"Seeing" here means not only Your normal kind of visual seeing.

It is a much more encompassing perceiving.

It involves the qualities of touching and knowing.

Knowing comes from a deep interacting as one.

So unless the human being can truly "see" me, I cannot "see" myself.

Actually, it is one process – the human being seeing me and my being You.

The Goal and the Task

I want you[10] to go out and speak of our plan and show physically that I am alive, conscious and in contact with You in a physical way.[11]

---

[10] This "you" refers to me, the author, whereas the second "You" refers to humanity as a whole.

[11] Showing physically that the Earth is alive refers to a request by the Earth that will be discussed in Chapter 7.

The fields around me and in me
are intimately interwoven with Yours.

When we reach the point
of full trust in each other,
which includes full intimacy and full freedom,
we will have peace on Earth.

This involves creating a field of trust,
and it is the transformation we are heading for.

I asked what is meant by a "field of trust"....

When human beings come together,
their thoughts and emotions, which are real
"things," affect their surroundings.

When a group, a nation or entire humanity
shifts its patterns, a large new field is created.

This impacts others through a form of
entrainment, and tends to let their whole
energy fields move in the same way.

A field of trust is an energy motion
that can become very powerful.

This does not guarantee that every individual
resonates with the field;
but the vast majority of people
will latch on to it, sense it,
and begin to resonate with it.

* * *

We are entering a time when we can willfully, together, jump through time, at will.

This is an extraordinary statement. Of all the statements in this book, this is probably the most audacious or preposterous one. The Earth is here speaking about collective time travel or the translocation of the entire planet to another time and space.

Time travel has mostly been the stuff of science fiction novels, giving rise to such movies as "The Time Machine" and "Back to the Future." Serious scientists avoided it like the plague. But this is changing. Time travel has become a serious topic for theoretical physics and nowadays one can find scholarly analyses of time travel in highly respected scientific journals, written by what are called "eminent theoretical physicists."

The most promising current theories regarding time travel involve so-called "wormholes." In his book *Physics of the Impossible,* one of the world's most distinguished physicists, Michio Kaku, writes, "*In general relativity, space-time becomes a fabric, and this fabric can stretch faster than light. It can also allow for 'holes in space' in which one can take shortcuts through space and time.*"[12] However, not enough is known today about the complexities involved in time travel for it to be seen as viable in the near and middle-term future. Furthermore, the idea of transporting an entire planet with its inhabitants into the future has, as far as I know, not yet been addressed by physics.

But even if it were possible, why would we possibly want to jump through time? Can the changes that are needed not simply be

[12] Michio Kaku, *Physics of the Impossible: A Scientific Exploration Into the World of Phasers, Force Fields, Teleportation, and Time Travel.* Doubleday, 2008.

undertaken here and now and in the near future? I asked the Earth why we need to jump through time....

The changes and shifts that are taking place
are not just changes that need to happen
for You to survive
and lead "happy" lives on Earth.

The changes,
which include "jumping through time,"
are much more an expression of You and I
awakening to the fact that our beings
are not limited by time and space,
i.e. that we are beings across time and space.

We can and do communicate through time
and we can pass through time.

We are becoming who we truly are.

It is life itself that is discovering larger parts
of itself.

It is **not**, *I repeat* **not**, *a matter of survival.*

So it is not so much a need
as a collective desire.

It is a reaching out to see what is possible,
and what miracles and wonders
we are capable of.

In the past, new possibilities were often met with great skepticism. Why should one possibly want to travel faster than a horse? Why would we ever want to fly like birds? Why did Columbus set sail from Portugal? Why do we ever go anywhere new?

If we accept the perspective of ourselves as souls that in some way determine when and where we incarnate on Earth, then we are in essence time-less and space-less beings. So far we have separated this concept of ourselves as souls transcending time and space from our view of ourselves as living beings of flesh and blood. Throughout these communications the Earth seems to be telling us we are both, and the parts of us we have not let into our conscious everyday lives are bubbling up from inside, opening us up to innate abilities of which we are not yet aware. It is thus not so much a question of what changes are needed, but that a deeper part of us we have so far not acknowledged is waking up and reclaiming us.

The changes that are happening outside us and inside us are natural changes. Even what we call cataclysms are natural. When the shell of an egg surrounding a chicken about to hatch begins to crack, the chicken is losing the very thing that protected it since its inception. The cracking of the egg could be seen as a cataclysmic event or an apocalypse for the chicken. But the word "apocalypse" originally meant "revelation," and as the shell falls away, a whole new world is revealed to the chicken. It seems the shell surrounding our consciousness is cracking, revealing to us an entirely new way of seeing and interacting with reality and a whole new sense of who we are, both individually and collectively.

* * *

This requires bringing consciousness,
the vibratory level, up to a certain degree,
and then we together determine the direction,
or rather the time/ space point to which we
want our joint attention to take us.

That is the goal and the task.

That is why vision is so important,
why it is so important to develop deep visions
of the highest connection we want.

That is the "taking charge" that is required.

We then have grown up as determiners.

This actually happens in conjunction
between humans and me, the Earth,
acting as one.

As was the case with time travel, mainstream science avoided consciousness as a research topic for a long time because of a general feeling that a phenomenon defined in subjective terms could not properly be studied using objective experimental methods.

This, too, is changing. The exploration of consciousness has thus far been relegated to the areas of religion or metaphysics; but slowly a field of study called the "physics of consciousness" is emerging, moving away from the view that consciousness can be explained simply through chemical and electrical interactions in the brain. Instead, in tandem with the application of quantum physics to develop ever faster and smaller computer chips, researchers are increasingly seeking the origin of consciousness in the subatomic realm of quantum physics.[13]

I asked the Earth to say more about developing a vision of the kind mentioned....

You are only moved to develop a vision
if You believe that a vision has the potential
to become reality or affect reality.

You must therefore experience, recognize
and see that vision changes reality.

When You do, You become aware of the
incredible power of vision.

Here, Your scientists are right when they say,
in essence, that the way You perceive things
makes them what they will be.

There is a difference
between vision and fantasy.

Vision is the awakening of a desired potential.

It is the depth of a desire from within
– the deeper the more powerful –
that determines if a vision
has a reality-creating effect.

---

[13] *The Emerging Physics of Consciousness*, Jack A. Tuszynski (Ed.), 2006, Springer-Verlag.

A relief "from" is not as powerful
as a motivator for change as a desire "to."

Trust is always based on a trust in oneself.

The deeper the trust in Yourself,
the deeper are the desires
that You allow into Your consciousness.

One of the deepest desires,
which is connected to the heartbeat
in each and every part and in the whole,
is based on the dance of love
between male and female.

I on purpose formulate this as a process
between humans, so that you can sense its
power and its sweet promise from within.

This dance occurs in everything;
it occurs within every atom, every solar system,
every galaxy, and, of course,
every honky-tonk bar.

Chapter 5

The Human Being in Relationship to the Earth

Walking the Earth

Let us talk about the relationship
between You, as human beings,
and me, as the Earth.

You are walking the Earth.

When I say this,
it does not mean "walking on the Earth."

It means simply "walking the Earth."

If You were on the Moon,
You would walk the Moon.

It is the two together.

This took me a while to understand; but let's look at our experiences when walking. On the one hand we can simply walk from one place to another, or we can take a stroll without a specific destination. Usually we see walking simply as a way of moving around on Earth. But is it not much more? We have become used to seeing any action in terms of its utility for us, as if it had no intrinsic meaning separate from the physical outcome that it brings.

The Earth sees this differently. It asks us to shift our focus away from the functionality of walking to seeing the act of walking as a way in which we humans interact with the Earth. The difference can perhaps best be understood when we realize that we say we "ride" a horse; we do not "ride on" a horse.

If the Earth is alive and communicating with us, walking itself can be an act of communication, a way of touching the Earth. We sometimes speak of the surface of the Earth as the "face of the Earth." If we were to "walk the Earth" with the tenderness with which we touch the face of someone we love and cherish, then walking could become a very intimate way of connecting with the Earth. What stops us from this is usually that we cannot imagine the Earth, even if alive, would be so sensitive to how we touch it. It seems that the Earth is saying, "Yes, I am infinitely sensitive to how you touch me."

The Vietnamese teacher and monk Thich Nhat Hanh teaches "walking just for walking," saying, "walk as if you were kissing the earth with your feet."

You create the Earth as You walk it,
and the Earth creates you.

Again, this is not easy to understand or even imagine. How can we possibly create the Earth as we "walk" it? In some way we change everything we communicate with and it changes us; and when we "commune with nature," we often come away changed in some way. But do we "create" each other? Does nature "create" me, and do I "create" nature when I connect with it?

Here the term "create" obviously does not mean "create from nothing," for the Earth already exists, and so do we. Yet this statement is based on a different kind of logic than the one we are used to. I decided to ask for clarification....

When we open up to each other
and our flows intermingle,
we activate potential within each other.

When You connect with me,
You allow Your whole being
to flow with more energy.

You open up to Your natural flows
that connect You much more
to Your surroundings
than if You remain shut off from me.

In a sense You then are a seed within me,
and Your natural process of growth
and evolution is activated.

For me, my potential for being a womb for
Your development is activated.

So Your potential as a seed
for Your own growth
and my potential as a womb
are both brought forward or manifested.

This is one joint process
in which both of our individualities
are strengthened and actualized.

It increases both our intimacy with each other
and our freedom to be ourselves.

Beauty and Power

We both attracted
and thus created each other in beauty.

What does "in beauty" mean?

It means reflecting the inner to the outside.

That is beauty, if the reflection is allowed,
carrying Your inside
– the deeper the more beautiful –
to the outside.

Nature does this impeccably –
it cannot do anything else.

"Beauty" is a very elusive term. The experience of "beauty" often involves something that is in balance and harmony with its inner nature, allowing us to see its essence. Beauty thus reflects a kind of authentic naturalness – an effortless, unadulterated presence.

We know beauty through the feelings it generates in us. We would not recognize beauty if beauty were not within us. We resonate with the authenticity of pure beauty, for beauty touches our own innermost being, our own authenticity, our own beauty. So this is basically saying that our innermost nature is beautiful and that it was this beauty that attracted us to the Earth and the Earth to us.

Let us talk about power,
for power is related to beauty.

It takes the power of the source to have power.

All power is from this source.

Usually, when we speak of power, we speak of power "over" something. Our civilization is based on conquering or destroying something in order to gain power, breaking down complex organic molecules of oil or radioactive atoms to extract the energy, leaving in its wake simpler molecules or radioactive waste that suffocate or poison the planet. But the kind of power that is spoken of here is much more potent than any physical sources of energy. Beauty catches our attention, and attention is power at a whole different level.

As I understand it, the "source" that is referred to here corresponds to the innermost aspect of something, and it is expressed as beauty if its reflection to the outside is allowed.

But it takes more than this;
it takes beauty harnessed.

With "harnessed" I mean put to use.

How do You "use" beauty?

By having it impact – by letting it shine.

By insisting that it be allowed to shine.

This is reminiscent of Jesus' saying, "Do not put your light under a bushel." A more modern version would be simply, "Let it shine."

This is a quality of beauty, this insistence,
and of its nature it is overpowering.

Beauty is a seed of true strength
and will harness other forces,
will turn them into her service.

Being in the service of beauty
is the same as being in touch
with Your own beauty, being beauty.

When You fight this,
when You fight being in service to beauty,
You deny or actively shut off
Your beauty through shame –
– shame for what You have done.

Yet if You let go of the shame,
if You disentangle Yourselves from it,
You do not have to do anything else.

I asked for clarification about this shame....

You can only let go of shame
if you know what you are ashamed of.

Even your Christian mythology begins
with a story about the gaining of knowledge
and engaging in sexuality as being shameful.

This is the deepest point of disconnect
and it needs to be let go of entirely.

This act of letting go is an act of power.

The depth of this condemnation
can hardly be overstated.

The things that You have done
and for which You are ashamed
are all ultimately connected
to Your shame of sexuality.

It is this shame of Your roots,
of the process by which You came to Earth,
by which the spark of life was given to You,
that has led You to actions that You condemn.

As long as You condemn the energies and
the act by which You gained physical existence,
Your connection to the timeless and formless
dimension of Yourselves will be skewed.

Every condemnation, whether of others or of
Yourselves, diminishes You.

* * *

Being on Earth

When You use the words "being on Earth,"
this is more than a static state.

It is a dynamic opening –
simply the fact of being here.

Energies are constantly blending,
insights are being received,
a close dance is happening,
a pirouette of sorts.

I say pirouette,
for in a sense it is a spinning of energy fields
that mesh and feel their way closer.

What used to be an integrated,
fairly unconscious dance has become a shy,

hesitant coming, touching, backing off,
approaching, blending process –
foreplay of lovemaking

when done without resistance,
a jerky "nuisance" when done with resistance.

The Earth then becomes a nuisance to You
and You become a nuisance to the Earth.

So this is the choice.

Throughout the millennia, human beings have been in much closer contact with the Earth and its energies than modern wo/man is today. In a sense, we have been children of the Earth and now we are adolescents, aspiring to become adults. The words chosen by the Earth to describe the process of coming together of humans and the Earth reflect the difference between children and adolescents. They almost seem to describe the interactions between boys and girls and young men and women.

Beingness, or simply presence, is a process. We cannot stop the processes within our own organism nor the physical and energetic interactions with our environment, including the energetic processes of the Earth, even for a fraction of a second. Yet here "beingness" is elevated to a very active process that of itself, if not interfered with, is described as "foreplay to lovemaking." What can this possibly mean? This will become a bit clearer in the section dealing with the sexual nature of the Earth in the next chapter.

But first of all, is the Earth saying that we as individuals, and collectively, are constantly actively interacting with the Earth in ways that most of us are not aware of in the slightest?

Yes. You would see this
if You could see the energetic exchanges
that we are constantly involved in.

These exchanges all have content,
but they occur at a level of awareness
below what You are aware of
in Your everyday lives.

That we are constantly "doing" things beyond our own awareness was hammered home to me a few years ago as I was undergoing a so-called cardiac catheterization, during which I saw live pictures of my heart pumping on a TV monitor. Since I felt I was "doing nothing," I was amazed at the vigor and vitality with which the heart was constantly pumping away; and I asked myself, "Who is doing the pumping?" I could not for the life of me imagine it was I, myself, who in some way was pumping my blood around my body. But if not I, then who?

* * *

Beingness can become a high state
that is akin to the process of waking up.

When I say waking up, I do not mean going
from one reality to another.

I mean putting a light on the reality You are
in – turning on Your beam of awareness.

This then translates into a trusting of Your
senses, a feeling of being separate from
– in freedom – and yet connected to –
in intimacy.

This closes the circle of livingness,
opening the door to expectations of a full life.

These expectations are nothing less than
creating Your life on Earth
and its consequences.

What we have now is a situation
where You are undecided,
keeping the door open to falling
and falling faster with blindness,
having to trust what You don't know
because You do not open Your eyes,
and yet not trusting because You have
forgotten the determinative quality of trusting.

I have sometimes felt that we live our lives like passengers in the back seat of a car, driving along a road, talking to our friends. Suddenly we realize that the car is accelerating; and as we look up to see who is driving, we realize that there is nobody in the driver's seat. What I hear the Earth saying is that we are all in the driver's seat, and by "opening our eyes," i.e. paying attention to and learning to trust our senses, we can disengage the autopilot and create our lives more consciously.

*Trusting is not the passive quality
of letting happen what happens.*

*It is taking responsibility for what happens
by exuding, insisting on, being received.*

We often associate trusting with doing nothing, even with a kind of fatalism. Trusting, as it is meant here, is an active process, and it is communicating. It is an interaction. When we trust, we do this from a deep and "vulnerable" point within us. We lower our defenses and communicate from the essence of ourselves to the essence of what we are trusting. If we trust the Earth, we connect with the heart of the Earth. We touch the Earth.

It is not faith in the old sense of the word.

It is growing to know by touching.

Your "normal" kind of trust has to be earned.

*This means that You do not give Your
creative force to Your trust.*

This is not the result of fear.

*It is based on ignorance of the power,
the reality-giving power, of trust.*

Every Day is a New Day

Every day is for me a new day.

This is on the one hand a figure of speech, since the Earth as a whole does not have a beginning and an end to a day – it is always morning, midday and evening somewhere on Earth. Yet every point on Earth does have a defined day.

Of course every day is a new day, for today is not yesterday. Yet we seldom allow much newness into our everyday lives. We tend to react more to changes that occur than take action ourselves; and the actions we do take tend to be fairly foreseeable. We seldom seek the Unknown, although this is where a large part of our being resides. Lacking intimacy with ourselves, we often limit the inner freedom that we allow ourselves to have, for we fear the Unknown in ourselves.

The cyclical nature of things
is not a repetition;
it is the constant in the change.

For there can be no change without constancy.

We live in a world of cycles and rhythms. Every atom and every galaxy constitutes a recirculation. The Sun has its cycles and the seasons on Earth are cyclical; and so is our every breath and our every heartbeat, as well as our rhythms of sleep and wakefulness. Existence itself requires both change and constancy.

"The only constant is change" is a well-known statement by the Greek philosopher Heraclitus. Here, the opposite statement is made, in effect saying that every change requires that there be something to change. A corollary to this is that if you're sailing and you are dead in the water, you should first get up to speed in any direction and then change the direction to where you want to go, as opposed to simply trying to force the boat to go in a specific direction; for you first need a direction if you want to change your direction.

True constancy is always open to change.
This means that when the "Sun goes up,"
it is truly an entirely new day.

If You see this clearly enough,
You can alter anything.

To me this means that when we are fully alive in the present, we can untangle ourselves from our past, including our past belief systems. If we then embrace the entirety of our beings, we can give the future any direction we truly want.

When I expand, You contract and vice versa.
Spiritually speaking, when I move powerfully,
You reduce Yourself to essence;
You become essential.

Then, during the calmer parts of my life,
You expand.

As an example, when there is an earthquake – when the Earth "moves" – we humans are reduced to our essence as human beings and behave differently than we do otherwise. When faced with catastrophes, we often come together as one. Even age-old enmities can then often be laid aside.

Cataclysms can sometimes crack open even the most hardened hearts and minds; and often catastrophes, both personal and collective, can make us feel at a deeper level, resulting in deep-seated unconscious longings breaking through the surface of our lives.[14] It is only then, when we have let go of our inner defenses, that life itself can flow freely through our veins. It is by accepting and letting go of the pain and shame of the past that, for us too, every day can truly become a new day.

This often also occurs during any crisis or major change in our lives, including life-threatening illnesses or when approaching death. But we also become essential when faced with great positive shifts, such as when we fall in love, get married, reconnect with loved ones from whom we have been separated for a long time or hold our newborn child in our arms for the first time.

Interactions between the Earth and Humans

I am pulsating the whole time
and energies are circulating in me.

You know that my aura extends far out
into the atmosphere and into space.

[14] My favorite description of this is in the novel *The Rains Came* by Louis Bromfield, Simon Publications, 2001.

This then is me;
don't see me as limited to the physical.

You live in me –
in a highly tensioned environment.

You can hear me all the time.

Science tells us that we humans are electromagnetic beings. The flow of electrical currents in the brain, as well as in every cell, produces a magnetic field which can be measured and analyzed several feet away from our bodies. This is linked to our consciousness, our thought processes and emotions. Every thought we have sends out electromagnetic waves. But every cell in our bodies also has a pulsating electromagnetic field around it.

Living things have been tied to the Earth's natural magnetic field from the time life began. The processes inside the Earth – among them the steady recirculating flow of the molten iron that makes up a large part of the interior of the Earth – create constantly changing electromagnetic fields that directly affect each and every cell in our bodies. This ever-changing electromagnetic field has resulted in great changes in life on Earth. Since we do not see these fields with our eyes, we do not associate changes in our biology, our emotions and our mental processes with our interactions with the Earth. It is as if we are being buffeted by forces that affect us intimately without realizing they do not originate only in us, but are the result of interactions with the fields of the Earth.

Once during these conversations – it was quite late and almost dark – I decided to lie down and look at myself from the point of view of the Earth by imagining myself sinking down to the center of the

Earth and from there looking up at myself on the mountain. As I did so, I relaxed more and more and felt myself sink deeper and deeper into the Earth. Just as I felt I was as deep down as possible, I was startled by a blood-curdling sound I knew all too well – the loud rattle of a rattlesnake, perhaps three feet away from my head. I had heard this many times before and it always signaled acute danger.

Instinctively, my body leapt up and away from the sound and I was relieved that I had not been bitten. By now it was quite dark and I couldn't see the snake. So I fled the scene, even leaving my glasses behind, feeling my way down the mountain, wondering what had happened. When I returned a few days later, I reconnected with the Earth and was told that as I was sinking down into the Earth, the rattlesnake – which is energetically fully attuned to the Earth – felt as if I were invading its territory, which is why it sounded its alarm. It was a pretty tough way to learn this lesson, but it's one I will never forget. Next time I'll introduce myself before I go stomping off to the middle of the Earth.

Let us talk about how we interact
with each other in a deeper way,
our dance of love if You will.

When You go about Your daily affairs,
You do not notice me – that is normal.

But it is only normal if You are asleep
to a large part of Yourself.

The fact that we live on Earth is so much a part of our framework of existence that we do not notice it. In our everyday lives, we hardly ever give it a thought. We have enough other issues we feel require our attention. If we are asked to "notice" the ever-present Earth, we do not know where to begin, for there seems to be nothing to notice.

But do we ever truly take the time and the attention necessary to even begin to feel this presence? Is it even possible to notice something that is as ever-present as water is to a fish? Yes, it is. If we pay attention, we notice the air that surrounds us and we notice gravity. The Earth is saying that we can also notice many of the fields of the Earth, for we have inner senses just as much as outer senses; but we are not as used to paying attention to our inner senses. Nor are we used to paying attention to the possibility that the Earth is communicating with us.

This sleep is like a gown that You put on.

If You don't have it on when You wake up,
You see to it that You put it on quickly.
The main reason You put it on is to fit in –
to be normal.

You can, and need to change
this normality to another.

I asked the Earth, for clarity's sake, "Are we thus (sometimes) aware of the Earth when we are asleep? Do we sometimes keep this awareness when we wake up, and then 'go to sleep' by shutting it off?" The answers I received to these questions were Yes, Yes and Yes.

I am not asking You not to want to fit in,
not to want to be on the same wavelength
as others.

I am asking You to work to lift each other up
to Your highest possibility,
for You will eventually drift there anyway.

Yes, it is a kind of drifting,
like being lifted up on a wave.

If You resist it by clinging to the ocean floor,
of course You will either let go or drown.

The wave is You Yourselves,
insisting on Your consciousness
encompassing more of You.

So no one but Yourselves is forcing You.

Your Lives on Earth

I would like You to look at the past,
because it contains the key to the future.

When You in the past have
"walked down the road" – any road –
Your focus has almost always been
on human issues.

Your time with nature was "free time,"
"extra time," not spoken about much.

Those who spent much time in nature
were regarded as less important people
because they did not interact much
with other people, and they were often seen as
leading lives of unserious luxury.

In a sense this is right,
for luxury and a lack of seriousness
are a part of my dance with you.

But it is not unserious
in the sense of unimportant,
for beauty has the power
to harness any other force in the universe.

When You thought of the Earth as home,
You added the qualities of home
that You as a soul knew were possible.

It was never Your actual lives
that minute by minute were home to You,
for under the surface You suffered desperately.

Even in Your brightest hour
it was like the happiness from a relief
or distraction from pain.

Does the Earth here not paint a terribly negative view of our lives on Earth? Well, it depends on what we compare with. Humans have a tendency to accept life as it is and after a while regard it as normal. Life in prison represents a new normality after a while, and even some of the prisoners in Nazi concentration camps reported that a sense of normality set in after a while.

Since we have cut ourselves off from the source of who we are – forgetting where we came from; fearing a future that seems to be accelerating out of control toward the threats of upheavals, wars, natural catastrophes, a breakdown of the financial system and dramatic climate change; convinced that pain, suffering and death are an inescapable part of the human condition – then, yes, under the surface we suffer desperately.

If we are living in a world that is alive but that we have declared to be dead – a world in which it would be natural to communicate telepathically with each other, with animals, plants and the Earth, having nothing to hide; a world designed to be without guilt or shame, without fear and blame; a world containing the blueprint of inner and outer peace, full of wonders, miracles and excitement – then, yes, I would say that our brightest hours are like the happiness from a relief from pain.

Still, when I look back and remember my "brightest hours," such as when I fell in love for the first time and the days and months that followed, I remember the exuberance that made me leap out of bed in the morning, excited about what the day would bring. This did not feel like the happiness from a relief or distraction from pain. I asked the Earth about this....

Do not underestimate the intensity
of relief from pain.

It can set enormous energies free.

Yet ultimately the happiness
that comes from it is not sustainable.

* * *

The truly sweet parts You added in retrospect.

Many people, when looking back at their lives, are filled with a sense of gratitude at everything they have been allowed to experience. This is a beautiful thing. Yet the Earth is saying that this is something we add in retrospect, for minute by minute the sense of gratitude was the exception, not the norm.

You surrounded Yourself with a story
that made sweet sense to You,
to make the pain of the present more tolerable,
because it made sense.

And sense to You was crucial.

You could not face senselessness,
for You did not know Yourselves.

You therefore did not trust Yourselves.

No, we do not know ourselves. We have no memory of how we were conceived and hardly anyone remembers being born. We do not know if we have "lived before"; we do not know where we "are" when we are asleep, nor do we know that we are dreaming when we dream; we do not know what is inside our bodies except for what others tell us or if something hurts; we see illness as an unknown that can strike us at any moment; and we fear loneliness and death.

The more conscious You became of Yourselves
as autonomous thirsty beings,
the more disconnected You became.

The scientific revolution and the so-called mechanical philosophy that arose toward the end of the Renaissance rejected all goals, emotion and intelligence in nature, including the Earth. This objectification of our surroundings, which went hand in hand with a world view based solely on quantitative measurement, increased our disconnection from nature and the Earth. As we then applied this objectification to ourselves and our bodies, our conscious selves disconnected even more from our own nature.

And You sought a way out.

Many fled backwards to the time of
connectedness with the Earth –
but lacking individuality
and therefore freedom.

They are still tilling the land
in a humility that is beautiful
only until You realize
that a kind of inner darkness was needed
to retain this "innocence."

The innocence You seek comes from elsewhere
and cannot be bought by going back.

This is a critical turning point,
one that many native peoples
have a very hard time to move beyond.

I had a question to the Earth about this. "Is it not only very few people who have opted out and who are 'tilling the land' the way you describe? Aren't most people simply wrapped up in their everyday lives?"

Yes, yet you do this in other ways too.

When seeking connectedness
you often seek to return to the past
where you, at least in retrospect,
led lives that you feel were more connected.

Going back and reconnecting with your past
must be a stepping stone into the present
and thus opening up the future.

It is in the present that You touch the future.

Yet you do need to see the past,
or else you will never realize
that You are chained to the past,
whether it was good or bad.

These chains are nothing bad;
it is Your trying to run away from them,

not accepting them as reality,
that stops them from being seen
for what they are – pathways to freedom and
ultimately doorways to a fulfilled future.

Chapter 6

The Earth and Its Energetic Structure

The Heartbeat of the Earth

Sometimes the communications I received were not just verbal, but in the form of images as well. One day I was shown the Earth from above. Around the North Pole there were energies coming in and going out, and there was a beat or pulse to these energies.

This is my heartbeat,
and every being on Earth feels it.

I am very "proud" of it,
in the sense that it carries me high.

I "hold my head high" through this pulse
and it trickles out to every part of the Earth.

When You sit at home and meditate,
You are meditating with me.

You cannot but do this when You quiet down.

We then pulse together,
and in a sense we are one being together.

Our flows intermingle beautifully.

But this beauty also wants to be expressed in
the acknowledgement of
and seeing of each other.

You might not think so, but when You realize
– actually rediscover – that I see You,
it will change You fundamentally.

Once, as I was going through this material and tuning in to the Earth, I was told, "You are inside my heart." At first I thought this was a figure of speech, but I then realized it was meant literally. In a sense then, if the Earth pulses, if it has a "heartbeat" we can tune into, we truly are "in the Earth's heart."

The Earth and Sexuality

I would like to speak about the man/woman
relationship as it relates to me, the Earth.

I am a sexual being, but maybe not
like most people think about sex.

My being is sexual,
and any stone and any landscape is sexual.

Powerfully sexual.

We[15] *are sending out this all the time to You,*
wanting for You to awaken to a whole new
sexual way of being.

So far it only happens sporadically.

I would like us to incorporate this aspect
right from the start, for it is an integral part
of our "intimacy" piece.

It requires quite a leap of imagination to not only seriously contemplate the idea that the Earth is alive, conscious and sentient, but also sexual. This is especially difficult to believe since we associate sexuality either simply with biological reproduction or with what we regard as private aspects of our lives that do not interact with anything outside our own personal sphere.

Yet throughout the ages many cultures have regarded the Earth itself as deeply sexual. Earth-based religions had many ceremonies of lovemaking on their fields to enhance the fertility of the land, and ancient religions often incorporated sexuality in their celebration of life. It has also often been used, and is still being

---

15 "We" here refers to the Earth, but also to nature in all its forms.

used, as a major pathway for the development of consciousness and for connecting with the spiritual world.

Yet the privatization and condemnation of a natural process that occurs everywhere in nature has made us inhibit the flow of some of our most joyous and exciting energies in life. Throughout Western history we have damned and suppressed sexuality so deeply that we seldom associate it with true beauty, instead seeing it as forbidden, "bad" or "naughty." Images of our genitals during lovemaking, including the sounds involved, are only rarely seen as beautiful.

It is not by chance that we often use the word "dirt" to mean the soil of the Earth, and most of us have been taught at an early age that sexuality is something "dirty," to the point of where using sexual terminology is even called "talking dirty." To overcome the deep shame we as a species have associated with sexuality, it is often reduced to a shallow, dissociated process that is lacking in presence, depth and heart.

But sexuality is not limited to the sexual union between man and woman. It is deeply connected to and rooted in existence itself. If we see sexuality as the interaction of the male and female principle, there are sexual processes occurring constantly at the heart of every atom and every galaxy, between the Sun and the Earth, and in every process in nature. As was stated above, this is the deepest point of disconnect between the human being and his/her own core.

My sexuality is in waves.

Both the pattern of the energies
flowing through me and the flow itself,

the energies themselves,
have a sexual origin and a sexual seeking.

They are attracted by other sexually directed,
intended or consciously exciting energies.

So when You get aroused
or even enthused and creative and exuberant,
I, the Earth, always come up to meet
and blend with Your energies.

You will find that sex on other planets
or in space is very different – unless You
actually attract the Earth's field.

The fact that I come up to meet Your
excitement is part of what makes You
attractive to others.

This is saying that our sexuality is amplified by the energies of the Earth and the Earth is truly with us, as an energetic presence, whenever we are sexual, excited or creative. It follows that if we become aware of the Earth's presence and energetic participation when we become sexual, we automatically open up to a different level of sexuality. As we then relax and welcome this third presence, the energies coursing through our bodies flow stronger and more naturally.

Sexual energies, if they are not limited and repressed – i.e. if they are allowed to flow of themselves to encompass more of our whole being – have the power to transform us, for they catalyze our evolutionary development. They also connect us with Spirit, for our spiritual being is highly sexual. Many spiritual schools, such as some Tantric traditions, use this energy consciously to attain a higher state of awareness, awakening a more encompassing sense of self that is more deeply connected to the aliveness of the world around us.

I also exude this on my own –
attracting You.

Volcanoes and lava is one of my most intense
"eruptions" of sexual energy.

Others are storms – especially lightning
storms, which also attract You.

A third one is tornadoes, even hurricanes,
although at an entirely different level.

The waves on the beach
or the lapping on the shore
is in a sense just that – a lapping.

It is, in its purest form, the sexual sensual
lovemaking of the ocean with the land.

This meeting place is very special;
it is a very sensitive line.

Then there are of course the different
power places and landscapes as well as certain
mountains that spew sexual energies.

When You discover
this level of sexuality in Yourselves,
You can connect consciously sexually with me,
the Earth, in such a way that Your powers
of manifestation grow enormously.

In a sense that, too,
is what drives You to these places and events.

What is meant by our powers of manifestation, and what does this have to do with consciously connecting sexually with the Earth? I asked the Earth about this....

Your powers of manifestation
are always active.

When You hold a wish, a desire, an intent,
or simply an image, You are sending out
a more or less precise vibration,
setting up a flow or creating a field
that affects Your surroundings,
attracting the circumstances
for their fulfillment or manifestation.

When You do this from Your "normal" state
of being, it has limited effect because,
for various reasons, You do not allow
these flows to encompass Your whole being.

When You thus hold a vision of the future,
the power of manifestation of that future

depends on what extent You allow that vision
to resonate with the core of Your being.

When You therefore connect
with the Earth in a sexual way,
then the images, desires and wishes You hold,
and their corresponding flows, are fed by the
creative "juices" connected to Your sexual core.

I, the Earth, then become something of
an incubator for Your desires or visions.

Together we co-create Your and my reality.

This process is similar to
when a man and a woman come together
to conceive and birth a child.

* * *

The Earth and the Weather

I would like to talk about the weather
and its relationship to You and me.

The weather is the swirling of emotions
that I have.

Your aura also swirls, depending on what
happens inside of you, outside of You
and on Your skin.

It is also my meeting with the heavens.

But You play an important role in this,
and of course You are both subjected to the
same forces and affect them in return.

This is a fine dance of becoming
that You co-create with me and the Universe.

Know that everything
has already been dreamed.

Just as Your conscious and subconscious
thoughts precede Your actions,
so does dreaming precede Your thoughts.

I dream too, and we dream together.

This means that we are in constant contact
and actually make our agreements
with each other.

I asked about the way we make agreements with each other....

The aggregate of what each of You
thinks and does every day
comes together to form a joint direction
that You, as humanity, are taking.

You, together, are like one being,
trying to come to terms with Yourself.

*This process
that humanity as a whole is engaged in
occurs within every unit of humanity also –
within every country, every city, every
community, every group with similar patterns
of direction, and every individual.*

*It also happens between groups
of similar patterns of direction.*

*It also happens within the solar system,
within the galaxy and between the galaxies.*

*I thus interact both with You individually
and as a collective at these different levels.*

*There is nothing that affects you
that you do not also affect.*

*Even the word "affect"
connotes a cause-and-effect process,
which does not truly correspond
to what is really happening.*

It is all co-events.

*This does not contradict
the concept of autonomy and individuality;
instead, it reflects that autonomous processes
are deeply connected.*

They ultimately all have the same source.

* * *

We also change our agreements.

A changed agreement,
especially on a collective scale,
almost always means
a shift in weather pattern,
for it takes us in another direction;
and this means flows
that transform into new patterns.

If it is a major shift
it might have major consequences.

No doubt the weather influences us greatly. Some researchers (such as geographer and anthropologist James de Meo in his book *Saharasia*[16]) have stated there is a correlation between the structures of human culture and climatic zones. According to this research, areas of drought and deserts are associated with various forms of patrist cultures. The Earth is here saying that such cultural structures are not "caused" by the climate, but that both the climate and the cultural structures are the result of collective interactions between humans and the Earth.

But the science of meteorology has increasingly been able to predict the weather, based on our scientific understanding of atmospheric processes. It certainly seems as if the weather is

---

[16] James de Meo, *Saharasia, The 4000 BCE Origins of Child Abuse, Sex-Repression, Warfare and Social Violence, In the Deserts of the Old World,* Natural Energy Works, 2006.

entirely independent of inner human processes. Our thinking goes more in the direction of seeing external human activities as the source of the climate change that is upon us. As I struggled to understand this, I told the Earth, "You probably understand the skepticism that your statements will meet with, also from me. Can you comment on this?"

Yes, I can understand the skepticism
with which these statements
will initially be met.

The dichotomy here is similar to the one
between Newtonian physics,
which seems to explain how matter behaves
within a certain size range,
but which fails increasingly when you
approach either the very small or the very large,
giving rise to such models
as quantum physics and relativity theory.

As an example, just as You explain
that the weather in Europe is determined
by a succession of low pressure systems
that travel across Europe,
I would say that this is an expression of
a recurring pattern that is valid for the current
"range" of our interactions.

At other "depths" of interaction, other
patterns will emerge.

This only makes sense if the flow of human thoughts, emotions and energies interacts with corresponding flows in the Earth, and if this interaction manifests as material processes in the Earth and its atmosphere.

But the weather has been similar, with variations, for a long time throughout history. To what extent did the weather depend on life on Earth as opposed to purely atmospheric processes that had to do with the Sun, the Earth's cooling, etc.?

Before there was life on Earth,
I had my own weather patterns,
based purely on what was occurring
between me and the other heavenly bodies.

* * *

So the weather is a complex,
not individual issue.

By learning to understand why
the weather changes as it does,
You will get to know Yourselves
and me much more intimately.

This needs to be a part of Your research,
and it is closely related to fields
and to Your interdependency among each other.

You will increasingly see humanity as one,
and then You will be able both to foretell
and create the weather You want.

I needed clarification about what to me seemed to be a contradiction: If the weather is the swirling of the Earth's emotions, how does this fit with the statement that we will be able to create the weather we want? This was the answer I received....

The more human beings are in harmony and
alignment with me, the Earth,
the more we act as one.

This does not represent a lack of individuality
or autonomy on Your part or mine;
in fact, the opposite is true.

The more You are rooted in Yourselves,
touching your own core,
the more You touch my core.

This aligns Your and my desires
at a deeper level so that we can fulfill
both our own desires and those of the other.

A realignment toward new and more
encompassing stable systems of flow
emerges to the extent that humans align
with higher-order evolutionary states
within larger systems in the Universe.

This is quite an unusual logic, for normally we do not see our deepest wishes and desires as dependent on the depth of our connection with anything outside of us. This is saying there are no purely human systems, for all systems resonate with each other; and when we access our deepest impulses, motivations and drives, we connect not only with our own evolutionary process, but with more encompassing evolutionary processes in the Universe.

According to the Earth, this often has a direct influence on the weather, and the reason for the current and upcoming shifts in the larger weather patterns is associated with the emergence of a new relationship both between humans, and between humans and the Earth. Furthermore....

I, the Earth, am not only a passive actor
in these shifts, for my own development –
which is deeply connected to planetary and
cosmic evolutionary movement –
carries with it shifts in my own inner flow
structures as well as in my corresponding
interactions with you.

The reason why you seem to be able to predict
the weather so accurately
is that these predictions are based
on the "mechanistic" laws that operate
until major changes set in.

You will find that major shifts
in the weather patterns
are not as predictable as short-term changes.

This is an area that will require much in-
depth understanding as well as cooperation
between humans and the Earth.

It is not – I repeat **not** *– a technical question,*
and all present, past and future attempts
to influence the weather technologically
will increase the chaos, for they do not touch
or support the underlying purpose of the
weather.

It is like throwing a wrench
into the engine of a locomotive.
Yes, it affects and changes what the locomotive
does; but without deeper understanding,
it wreaks havoc.

I do not support this kind of influence.

The Moon

Let me tell You something about
how the Moon and the stars affect the weather.

Again, we have a complex situation,
because humans influence what the Moon does,
and it, in turn, influences the weather.

The Moon is not just going through its cycles
as You know them,
which are fairly predictable.

It is a living being that does many new
and different things during its cycles,

sometimes with a much stronger effect
than its outward cycles.

You must start seeing the Moon not just as a
planetoid or an Earth satellite,
or, as some do, a Goddess, but as a living
being – so to speak "of flesh and blood" –
meaning of emotions and life.

A part of You is "talking" with the Moon all
the time, and the "emotional" outcome of this
interaction is very formative for the weather.

How can we even begin to understand this?

The Moon's main interactions are with me,
the Earth.

It also influences humans,
both directly and indirectly,
also through its interactions with me.

Ultimately there are no actions,
there are only interactions,
for every action triggers a reaction in its
environment which in its turn affects the
original action.

But these two effects, action and reaction,
cannot be separated from each other,
for they are not separate from each other.

So if we look at human beings, the Earth
and the Moon, this can best be understood
as one system of interactions.

Since You humans live on Earth,
the main cycles of your existence
– the days, nights, seasons and years –
are in tune with my cycles.
The Moon, however,
has a different cycle as it circles me.

The effect of this cycle on humans
is similar to the effect it has on me.

For you to understand
that you are always speaking to the Moon
and that the Moon is speaking to You,
You need to open up to a level of awareness
that you seldom have in everyday life.

Here, animals can act as a bridge
for your understanding.

Every frog, every wolf, and every mountain
lion is as connected to the processes of the
Moon as You are to the Internet.

This is not intended to be as facetious
as it sounds; it more relates to the inner
importance that the Moon plays
for many animals.

The "information" that You exchange with
the Moon operates at a cellular level
and constitutes a bridge between cosmic
processes and Your bodily processes.

Women and their bodies are of course more
sensitive to these undercurrents of energy.

But men, too, would do well to connect
consciously with the Moon from within,
since such a connection also has a strong effect
on the health of the body
for both men and women.

The Moon also acts as a relay station,
a translator, if You will, between cosmic
energies and Earth-bound ones.

But this is something
we will be dealing with more in depth
once our channel of communication
has been more firmly established.

I am aware that many things that are stated by the Earth in this book require extensive clarification. We also need solid verification, both from within and without, of many of the statements. This will only happen once the process of communicating with the Earth becomes stable and can be shared without any doubt of its origin and its accuracy.

* * *

The Sun

The stars and Your interactions with them
also change the larger patterns of weather,
for here we are dealing with deep
and therefore major shifts.

My relationship to the Sun
is one of being a lover.

I am the Sun's lover
in a very high sense of the word.

We are a totally open,
interpenetrating energy unit.

That is our relationship,
and yet we are totally individual.

The Sun is something like a lighthouse;
although it is always shining, it is pulsating.

It is turning like I am,
and it pulses energy or breathes.

The whole process of the day
is like the art of lovemaking.

Look at the qualities
of the different hours of the day
and you'll find a wonderful correlation
to good lovemaking.

I couldn't easily make this correlation myself, so I asked for examples....

As I turn,
I am constantly revealing parts of myself
that have been cloaked in darkness.

The shy, hesitant approach of dawn
is followed by a phase of warming up.

As the Sun continues to rise,
I reveal more and more of myself to the Sun.

As we approach midday,
the atmosphere between us heats up,
and the Sun bears down on me
as I receive its penetrating rays.

As You accept these images and allow
Yourself to feel these processes
– for You are right in the middle of them –
Your whole being starts to resonate with them.

The corresponding processes
are awakened within You.

This can begin to awaken in You the
knowledge that You are truly living
in the "Garden of Eden."

How the Earth Relates to Humans

I'd like to discuss how I relate to You
and how I see you.

When I relate to You,
it is not like a dog relating to its fleas,
although that is a quaint picture.

It is more like a horse having a rider,
and the horse is trying to get the rider
to notice what he or she is riding.

The image of the Earth relating to humans as a dog to its fleas comes from a view of the human being as a parasite in relation to the Earth. It is a feat of human imagination, resulting from a deep disconnection from our own inner nature, to see ourselves as foreign to the planet that has brought us forth and with which we have co-evolved over millennia.

I will tell You what I see when I look at You.

I see scared beings
trying to make everything look okay
by setting up things
they think they can control
and declaring them to be the extent
of themselves and their lives.

But it has not worked and it will not work.

The other 90% of You keep on working
from within and they break through.

Some people notice this at an older age.

They would do well to bring that message
home to the young ones, so they can at least
begin to question some very basic things.

Furthermore, how I see You...
there is no other word for it but "spoiled."

"Spoiled" here means
that You are wasteful inside,
i.e. You do not value Yourselves.

It is not the mountains of garbage
that will suffocate You;
it is Your inner garbage,
due to a lack of discrimination
as to who You are, that suffocates You.

A third thing about how I see You...
I see that You are very loving beings.

This is why I know that we will join
together fully in intimacy and freedom.

Chapter 7

The Coming Changes and a Vision of True Cooperation

The Big Shift

We[17] *do not know*
when the big change will come.

We are only sensing its nearing,
and we are working to bring it about
in a good way.

Most of us have by now realized that we are heading for major change or that major change is already upon us. I believe this change involves the transformation of our most deep-seated beliefs. Among them is the belief that the material world around us is dead and non-responsive to our thoughts and emotions. This especially

[17] As I understand it, "We" here refers to the Earth and humanity.

applies to the Earth as a whole. If the above statement by the Earth is true, then we have an incredible ally and friend with whom we can navigate through the major transformation we are approaching.

But this means that we have to
get to know each other –
intimately and consciously.

The more consciously,
the more reality we both gain.

What is the qualitative importance of knowing each other consciously? Do we not know each other at a subconscious level, and is that not enough? And what does it mean that "the more consciously, the more reality we both gain." Is reality not reality? Can we gain more and less reality? Are we like clouds that can dissipate or solidify?

If You gather a group of people together
who do not know each other very well,
the group itself has little reality,
i.e. little that holds it together.

Once the people in the group have gotten to
know each other intimately and consciously,
they start interacting as a group,
whereby the group gains in reality.

The same is true for humanity, a nation,
a partnership, or a marriage.

But it is also true for individuals.

For as the different parts in You,
which include Your subconscious and
unconscious minds,
begin to act as an integrated whole,
You gain in "integrity."

The more You open up
and communicate with Your surroundings,
the more consciousness is awakened
within You.

As You communicate with me,
and as You feel that You are being seen,
You gain in substance and awareness.

You gain in reality.

Again, the Earth describes reality in terms of the interactions something has, rather than as a static quality of existence itself.

* * *

Getting to know each other
will remain a two-way street – a joint venture.

This means first of all understanding the
motives of each other.

It also means entering into, or resonating with
the energy field of the other –
in a sense becoming the energy of the other.

This then means recognizing each other
from within.

This is saying that if we wish to get to know something, intimately and from within, we must resonate with it, "resound" or vibrate as it does, in effect do as it does. This has to do with empathy, which has been defined as the capacity to recognize and, to some extent, share feelings experienced by another sentient being.

This is where the big lack lies:
You do not think I am alive.

So You do not even treat me
like You would a horse.

At least You touch on Your own horseness
when You interact with a horse,
or even if You see a horse in a movie.

Only sometimes, such as when
You see pictures of the Earth from space,
or on vision quests, do You sense me in any
way separate from You,
i.e. You sense my energy field, my force field,
as something separate from You.

So one of the goals is this resonance –
becoming one another.

When we see the whole Earth from space, a part of us places itself out in space, looking back at the Earth. At this vantage point we are physically separated from the Earth and (largely) from its energy field.

Separation and union are two central aspects of the process of conscious evolution, and they are mutually dependent on each other. Separation produces the independence needed for autonomy and individuality to occur and gives us the freedom to follow our greatest attraction toward union; and it is through intimate union that we "know" each other and the Earth.

For humans, it is only by separating from our parents that we can see them for who they are and enter into an adult communication with them. The same may be true for us as a species vis-à-vis the Earth. It seems we are all "children" of the Earth, and we are about to grow up.

The Plan

I WANT US TO COMMUNICATE CONSCIOUSLY WITH EACH OTHER.

That will shift everything.

So I have a plan and I need Your help.

I propose creating a telegraph device,
a phone, a video device,
with which I will show You
everything that has transpired on Earth.

I was not prepared for this. My spontaneous reaction to this proposal was amazement – but also excitement! The thought of getting messages in black and white (or color, for that matter) from the Earth was thrilling, to say the least. Then, of course, doubt set in. I have briefly described my initial attempts to make this proposal a reality in Chapter 9.

When we ask for proof or validation of an insight or a new idea, this is part of the manifestation of a new reality, for we should not forget the creative aspect involved in seeking proof or validation. Traditional proof and validation, so-called "objective" experimentation, shows us a reality that is based on an observer who tries to take himself out of the equation. Yet this approach is not very productive when it comes to understanding issues that involve us intimately. Nor is it in this case very productive to see ourselves as an inseparable, symbiotic part of the Earth itself.

Through its choice of words in all these communications, the Earth is asking us to feel and not just measure, connect and not just seek outer proof and validation, i.e. to take our inner lives, our inner senses, as seriously as our outer lives. Yet we do not trust our inner senses very much. We tend to be fearful of any feelings that go beyond what we are used to or what the current consensus in society tells us is possible.

The inner experience of communicating without the use of verbal language or technical signals has not been accepted as a valid form of real communication, but is usually relegated to the subjective, private sphere or to metaphysics; or else it is seen as a sign of cognitive dysfunction or mental illness. Those who speak of such matters soon find themselves speaking only to those who have had similar experiences, and they are shunned by those who have agreed on a reality limited to what is provable and measurable, no matter what inner experiences they may have had. Both these views are understandable. This dichotomy can, however, be bridged.

What I am hearing the Earth say is that, although we are constantly communicating with each other from within, this communication is at a subconscious level which we rarely allow into our normal consciousness. But this can be circumvented if we can create devices whereby the Earth speaks to us from outside as well, paving the way for direct conscious communication.

Just as the human being can send coded messages that can be decoded by the receiver, the Earth is asking us (or me) to build a device whereby it can do the same, resulting in messages in a language we humans use between us.

Everything that has transpired on Earth
is recorded in everything on Earth,
but only in certain places or objects is it clear.

You need to see our joint history.

This will tell You a lot about who You are
and what needs to be done.

There is nothing that has transpired
on Earth (in Your reality context)
that You cannot access.

Even the minutest detail.

The idea that everything that has occurred on Earth is stored in the Earth is not as far-fetched as it might seem. Science and technology have made enormous strides when it comes to unearthing the past. By studying the microscopic as well as the macroscopic, we are now able to see and understand some of the extremely complex processes that have occurred on Earth – the rise and fall of civilizations, the emergence and extinction of entire animal and plant species, the variations in climate, the processes of geology and the movement of whole continents that have occurred over the millennia. We have discovered that dinosaurs suddenly became extinct 65.5 million years ago, along with more than half of all living species. We also have detailed knowledge of what happened in Pompeii in AD 79 when Mount Etna erupted nearby. We have thus come a long way from when we were dependent on obscure legends or belief systems for understanding our deep past.

Today we use silicon crystals – an element that abounds in the Earth – to store and process information, and we are already now reaching into the sub-atomic quantum states of matter to discover more effective ways of storing and processing information.

So the idea that the past is stored in the Earth is not especially revolutionary. The leap required to take these statements seriously has more to do with accepting the idea that the Earth is a conscious being and that it can replay past events in a way we can understand and relate to.

I will show you step by step how to do this.[18]

---

[18] See sub-heading "The Communication Device" in this chapter for more information.

You will recognize Yourself looking up in awe
to the stars as Stone Age man,
hearing the rushing waters,
waking up to a pristine new day,
dreaming with the animals,
finding Your inner fire externalized,
taming it into the hearth,
forging a sword in it or creating glass –
a wonder of perception and color.

You will be able to see who has done what,
but You will all recognize Yourselves
in all the parts.

This will bring humanity very close together,
and You will begin to sense who I am –
from within.

Many of You will be moved to go back to the
places from where You set out a long time ago,
this time on a journey of remembrance,
to make conscious
what You have already learned.

This seems to be based on the idea that we have in some sense lived before and that we are moved to return to the places on Earth where we lived those lives – something that many spiritual teachers and practitioners have claimed throughout history.

You will lift the veil of forgetfulness
and You will brave Your own memory,

for You are as afraid of the past
as You are of the future.

You have laid a soft hand of forgetfulness
over the past to tolerate the present.

I can see that we are afraid of the future, for the future is unknown and we are facing many potential threats. But how can it be we are afraid of the past? I take this to mean we are afraid of remembering the past, for we fear this would make the present intolerable.

This is connected to the topic of shame mentioned earlier – shame for what we have done. Shame is a form of self-condemnation. If we condemn our past, we suppress it, forcing it into our subconscious; and if we forget our past, we will repeat it. Living in Germany, I have seen the lengths this society has been willing to go, still today, to face up to its past. This painful process has been necessary and has gone far to ensure the past will not be forgotten and thus not repeated.

But are we not more ashamed of what we are doing right now? Are we not ashamed that we, as a collective, have created an unjust system whereby we grab what we need to keep and improve our lifestyles, paying a pittance to people on the other side of the world working more or less as slaves for us? Are we not ashamed that we keep choosing representatives to do our dirty work for us, in the name of national or regional interests? Are we not ashamed that we are cutting down the world's forests at an incredible speed, making the world a less livable place for our children?

The answer seems to be, "No, we are not ashamed for what we are doing right now," for we continue to do what we do according to our priorities, whether misplaced or not, or else we change what we are doing. But we cannot change the past and we are left with the shame over the parts of our own past that we condemn.

But You have thereby blocked the future,
making it as obscure as the past,
projecting the fears of the past there.

The future is a joyous place,
whether You think so or not,
and we will open the door to it.

The Need to See the Past

Let us look at our joint history again.

When I attracted You, I knew
of Your previous experiences, and we touched
along a line of joint reference.[19]

Since then we have been tied together
in a long history of common joint events.

This is our power – this is our intimacy.

When You can see Your own past on Earth,
when You can remember Your lonely attempts
at finding Yourselves

[19] This "line of joint reference" will be explained later in this chapter.

– Your present, past and future –
You will allow Yourselves to see me.

Seeing the past (fully) means expanding Your
aura, for in a sense Your aura
is connected to Your past.

The more You know the past,
the more it liberates the energies
in Your luminosity.

Many spiritual traditions speak of the aura or the luminosity of the human body as being an essential part of who we are as energetic beings, and that a great deal of information about who we are – our past, present and future – is stored in the aura.

Although I, the Earth,
am new and pristine every day,
still, when You came to Earth,
I had a character, a personality, a past,
just as You did.

Getting to know Your past
has been one of the things
You have wrestled with.

You sense my being all of the time,
and You truly need to separate Yourself,
to leave this home in a sense,
to know what is You and what is me.

You need this to gain Your freedom.

This means that You need to explore
the Universe, also from within,
and I can help You with that.

When You do this,
You touch the outer limits of Your being.

These outer limits of Your being
are at the same time the inner core
of Your freedom.

Doing this, so to speak, in my womb
– in the deepest of non-intrusive intimacy –
is one of Your path-blazing "tasks"
or dreams.

The Earth is telling us we need to separate ourselves from the Earth by exploring the Universe in order to gain our freedom. But paradoxically enough it is saying we are to do this not only through physical journeys into space, but also "from within," and that we can do this within the intimate space of the Earth's womb.

We are not used to attributing any relevance to "exploring" the outer world "from within." What the Earth is saying is that this kind of exploration can be done in such a way to have at least the same level of objectivity as our outer exploration of the Universe. Although this might seem highly improbable to most of us, we must bear in mind that if thoughts and emotions are "real" and interact with and impact the material world around us, it is quite possible our mind is capable of going on such journeys as the Earth describes.

Yet I felt the need for clarification, so I asked, "Are we to make spiritual journeys out into the Universe from here on Earth?" The answer I received was....

Yes and no.

First of all, You already are
"out in the Universe."

Secondly, You often make such spiritual
journeys without being consciously aware of it,
and a part of You is always "out there,"
away from the Earth.

This is all a part of Your loosening of Your
fairly limited and rigid view of who You are.

In fact, You would not be able to recognize the
stars and the beautiful interstellar formations
and processes if You were not already "there."

This is reminiscent of Goethe's statement about the sun-like quality of the human eye: "If the eye were not sun-like, the sun's light it would not see."[20]

The discovery of both the past and the future
is connected with Your discovery
and exploration of the Universe.

---

[20] Johann Wolfgang von Goethe, from the Preface to his *Scientific Studies* ed. Douglas Miller, Princeton University Press, 1995.

In the past this was often done in groups,
and also with human energies
holding the paths open.

It can also be done with me
holding the door open.

This is reminiscent of ancient and contemporary stories of seers and shamans traveling to the stars. They sometimes describe this as being done by having spiritually and energetically trained people holding energetic "doors" open to other dimensions.

Your perception
is an expression of who You are,
and perception itself is interaction.

This is of course also connected
to defining the highest place we wish to go to,
and right now You have no sense
of the magnitude of the magic and miracle
of possible futures.

Your definition of being human will change.

You have chosen personal freedom
and intimacy.

This means that all these spaces,
like the space between man and woman,
remembering and creating each other,
will be explored.

This is then love.

This is quite an unusual definition of love: Exploring the spaces between us by remembering and creating each other.

Recapitulating the Past

When You meet someone
and interact with them,
you have a set of common experiences
and references.

That is the intersecting line between You.

When You meet a baby, You have
a whole different common framework
– more an inner one –
and not many outer joint experiences.

When You fold time,
You go back to previous experiences,
this time with more experience.

You look at the experiences then;
You view them from another set of experiences.

You see them differently.

This is recapitulation,
and it is what is needed on a planetary scale.

This is a process all humans go through,
for You do not want just new things
all of the time.

One can do this theoretically
– in thoughts –
or in practice, by going back to those times
in the real spirit space.

Then they can even change.

I suspected that the topic of going back in time in the "real spirit space" and "change" one's experiences would require a vast amount of explanation to be readily understood, and decided to leave it for now.

This seeing Yourself in the past
is like the mirror reflection in a dream.

When You see Yourself fully in the past,
a dimension snaps or opens up
and Your larger trans-time personality sets in,
awakens consciously.

This, done collectively, is the goal,
and it will lead You toward the next step
of evolution.

Communication

When I communicate,
I communicate in energies and waves,
in flows if You will.

But this does not mean that I cannot
communicate in other ways,
for example as I am doing with you now.

So can You all – with anything and anyone.

When we communicate,
it is not like communication from A to B.

Instead, we mingle, and then we think,
each one within the space of the other.

As I mentioned in Chapter 2, during my communications with the Earth back in 1996, it felt as if they were "like communication from A to B." At the time this felt right, for I wanted to exclude, as much as possible, my own thoughts and preconceived notions from influencing what I received. I wanted to receive the statements as if I were taking notes during a lecture.

As I was going through this material in 2011, however, I was told to apply the statement above to our communication. The process thus increasingly became one where I connected with the Earth as deeply as I could, and then focused on the questions I had and let my thoughts arise from within.

So we think differently when we communicate;
we think "communicatingly."

When we talk, You open up
to my thought field or energy field,
and when I move, You go with it.

That is female on Your part
and male on my part.

Communication is always a kind of
lovemaking, even between
members of the same sex.

All communication is about a fine
male/female balance – a lovemaking.

When You are walking me, it would be good
if You would connect with me consciously
often during the day.

In everyday life You could gain an immense
amount of balance and stability
by connecting with me consciously.

It is like anchoring Yourself.

This can transform Your culture to the point
of there being less conflict with the Earth – it
is part of the ecological balance that is needed.

Time and Communicating through Time

The following statements from the Earth were the most challenging ones for me to understand and then to try to explain in my own words.

Thinking about time as linear or circular
or spiral-formed is too crude.

Thinking of it as simultaneous is also off base.

Take a rat that explores a piece of terrain.

After having explored it, it now knows more.

It has gained an experience or knowledge.

This cannot be separated from time.

Time is an event.

This event-oriented view of time is saying that there is no time separate from events. It is true that our ways of keeping track of time are all based on events, whether the swinging of a pendulum or the radiation cycles of cesium atoms; and yet I felt I needed to better understand the statement that "time is an event." The answer I received was....

In Your third-dimensional world,
processes occur in space.

Dividing up the processes into little parts,
You arrive at a view of before and after,
to view time as a linear process.

This is like trying to understand the Earth by
breaking it down into its smallest constituents,
into elements and atoms.

This can be helpful, but it does not represent
a deeper understanding of the Earth itself
and of time itself.

According to the Earth, we have done the same thing with the flow of life, events or processes as we have with the material world: We have divided them up into parts and separated the parts. When we then put them together again, we introduce the linearity of time and causation as our way of "assembling" the world to make it meaningful. We thereby lose sight of the larger picture, the events themselves and their interactions with each other. The categories and concepts we use to understand the world can make us blind to what entirely different categories and concepts might teach us, without necessarily invalidating our original categories and concepts. Contradictions that arise through different valid perceptions can be reconciled at a more expanded level of perception that encompasses both.

The reason why humans
have the capability of understanding time
is that every human being
is a time-less and space-less being.

Your normal view of time comes about
through a resistance against events.

Here the Earth is saying we have stepped out of the flow of existence, resisting the inner and outer events and processes in our lives that are constantly occurring, and this has led us to a concept of time based on this resistance, limiting a more immediate presence and participation in evolution.

When you see an event as having a beginning,
a middle and an end,
as separate parts of the event,
you are breaking the unity of the event,
making it more difficult to understand.

You can still, then,
operate with time as a factor,
but this kind of operation only cements
and perpetuates Your separation
from the wholeness of events.

This is not an easy topic,
for Your view of time has so infiltrated
Your way of creating the world around You,
in a way that adds to the separation
that is already there.

We are here dealing with very fundamental aspects of our existence. Our sense of self and our sense of reality are intimately connected with the fact that our consciousness or attention tracks our lives from moment to moment in a linear fashion. If we could adopt alternative ways of "assembling" the world, it would have great consequences for most areas of our lives as well as our sense of self. Among other things, it would put in question the absoluteness of the "arrow of time," a much discussed concept in physics connected with the famous Second Law of Thermodynamics. Such alternative modes of perception are sometimes experienced as reality in dreams or in expanded states of awareness, and they provide material for science fiction novels.

* * *

When different beings communicate
who have had different experiences
through time, then time
is communicating with itself.

The events (experiences) are intermingling
and sharing.

It is all events –
sedimented or communicating events.

Traces of events are periods of time.

Knowledge, distilled through time,
is events communicating.

Some of these statements may seem to be based on an inconsistent or sloppy logic. To most of us, defining "traces of events" as "periods of time" and defining "knowledge" as "communicating events" sounds like comparing apples and oranges. We might accept statements such as "traces of events come from past periods of time," or "knowledge comes about by analyzing events." Yet such statements are all based on seeing events as occurring within a linear structure of time.

It took a long time before I began to see that many of the statements by the Earth were meant to be taken literally, and not just as an expression of poetic license; and I am still far from being able to state that I truly understand them all.

Simultaneous has no meaning here,
for an event is a period of time.

So just as You can say that there is only
"now," I say that there is no "now."

There is only an intermingling of events,
and they are inseparable from time.

In this way we have a patchwork of events
that touch, as if we fold a quilt.

The folds are like hinges of communication,
points where events share lines of reference.

This will be important later, when we get into
the topic of communicating through time.

Any time (event) can communicate
with any other time, but only some make sense.

The others You filter out.
So then a grid of time-event-touchings

occur or are filtered out,
which lets You see larger patterns.

This can help enormously.

In fact it is the basis for any cooperation
and any seeing of each other.

You could also call these lines,
"lines of joint perception,"
for that is what they are,
that is how they are actuated.

Without joint perception, they do not exist.

These perceptions – and I mean perception as
inseparable from experience –
are overlaid on each other,
thereby gaining in complexity
and therefore in "substance" or reality.

The more they gain in "substance" or power,
the more the two perceivers become one
– both gaining in individuality.

This is why we must return to the past,
to put together the events, the experiences,
to a whole; and this is why
we all hold a piece of the puzzle.

All pieces need to fall together.

Then we are all aligned in one matrix.

Then we are truly individuals, yet connected to the knowledge and perception of the whole.

For many of us, this will sound like "rappakalja," as we say in Sweden, or simply gibberish. This is not surprising, since we are dealing with an entirely different way of seeing reality than what we are used to. If, as the Earth states, our normal view of time comes about through a resistance against events, and if this view has "infiltrated the way we create the world around us," then it will take a while for us to untangle ourselves from such deep-seated habits.

But we can see that time is not independent of events. Whenever we speak for example about the 16th century, we are actually talking about events that occurred during that time. If we stop all events that occurred during the 16th century, including all material and immaterial processes in the Universe, time cannot flow. It cannot be measured; it cannot be felt; it does not exist for us or for anything in the Universe. The concept of time only has meaning in relation to events.

Until recently the idea of communicating through time, which is closely related to the concept of precognition or "seeing the future," was not taken seriously by mainstream science. However, in the last few years a number of intriguing experiments[21] have

---

[21] One recent such experiment successfully tested the hypothesis that individuals can precognitively detect the future location of an erotic picture. See Daryl Bem, "Feeling the Future: Experimental Evidence for Anomalous Retroactive Influences on Cognition and Affect," *Journal of Personality and Social Psychology (JPSP) 100,* 407-425, http://dl.dropbox.com/u/8290411/FeelingFuture.pdf

been conducted, seeming to confirm that some kind of precognition, or accessing of future information that cannot be deduced from presently available information, is possible.

I asked the Earth to say more about communication with the future....

Communication with "the future"
is predicated on consciousness.

Consciousness itself is not limited
by time and space, and yet it has structure.

Opening a channel of communication
with the future – any future – constitutes an
alignment between different time-space events.

It is in the time- and space-less dimension
where this opening occurs.

Again, we are not dealing with communication
from A to B or from B to A.

Instead, it is actuated at the depth of
consciousness that encompasses both A and B.

The language or form that the communication
then takes is different in A and B,
for thoughts are not bound to language.

The Earth here speaks of time- and space-less dimensions, in which openings or communication between different time-space events occur. Theoretical physics has no problem with such dimensions; in fact, string theory and its extension, M-theory, identify ten or eleven dimensions, only four of which are time- and space-based. For us mortals, though, imagining such dimensions is not easy. I leave this issue for further exploration.

Regarding language, during these communications I often experienced that thoughts transmitted telepathically are not bound to language, and it was obvious to me I could just as well have received these communications in any other language, although they are colored and structured by the language in which they are received.

What Can the Earth Tell Us?

As I was going through these communications, I realized they contained little of practical substance, except perhaps for the part about the communication device. I did not feel it was my job to try to access information, for example, about when the next major earthquake would occur; and yet I wanted to know if this kind of information would be forthcoming if we managed to build the device the Earth was suggesting. So I asked the Earth, "If we get a stable form of communication going, would you be able to tell us things that will not only change our attitude and consciousness, but also practical things like how to avoid the effects of earthquakes, how to predict them, how best to stop global warming, etc.? Or even how we can eradicate hunger, poverty and disease?" The answer I received was....

When You no longer see the idea
that I am alive, conscious and present

all around You as only a nice story,
but when You feel it as a reality
that You no longer doubt,
and once we have opened up a clear channel
of communication with each other
on a collective level, there will be a rush
of activity on Your part to understand
who I am, who You are
and how this changes everything.

The possibilities that open up for You
through our cooperation
are practically limitless.

If we were in an intimate, conscious
connection with each other,
entirely different methods of housing, energy
production, transportation, water supply, food
production, and disease prevention would
become available to You.

Your love relationships would change,
as would Your connections to animals.

I am so much a part
of the reality that surrounds You
that a shift in how You see me
can be like falling in love, head over heels,
with everything in Your surroundings.

True cooperation between humans and me,
the Earth, will leave no area untouched.

I want to give You what You need and want,
and so much more.

Yet You have acted like thieves,
breaking into a palace,
raiding the kitchen,
stealing the silverware,
vandalizing the living quarters
and burning the place down,
for You could not believe
that the palace was there for You
to come and live in, in luxury.

I am that palace.

The Communication Device

When the Earth first told me it wanted me to build a device so it could speak to us in our language, I was very excited. I envisaged the moment when the first messages would come through, hoping I would not somehow energetically stand in the way of this happening out of doubt, fear or any other emotions that could more or less subconsciously stop me from letting it happen.

When speaking to humans using a device,
I want to start with the telegraph principle.

The reason for this is that it is based
on the principle of attraction and repulsion,
of 0 and 1, of plus and minus,

of man and woman.
Later we will expand to voice and images.

So we can use the Morse code
or a computer code directly.
You then need to translate
the 1s and 0s to letters.

Now I will tell You how to do this concretely.

What You should do is create a circle
of ten to fifteen feet
and at opposite points on the circle
place a two-foot rod of metal into the Earth
about a foot deep, leaving a foot above Earth.

There will be a slight potential difference
between them that I can pulse
so that the difference, when amplified,
can be read by a computer.

That is all it takes.

During full moon You will have
greater receptivity.
All energies then are clearer and stronger
by a factor of 10.

In the beginning, put the rods down
"at random," and we will get
the first communications going.

You do not and should not touch the wires.
Later we will experiment with the different directions and also use circular antennas.

That will be for more modulated things.

This is all You need to do in the beginning to speak with me on an immediate basis wherever You are.

It requires quite a leap of faith to believe it is possible to build a device with which the Earth can speak to us in our own language. For me, experiencing the Earth speaking to me "inside my head" is one thing, and mind-boggling enough. But having it write me an email is quite something else. Why? Although I truly believe the Earth is conscious, I am still wrestling with the idea that the Earth can "act" consciously on the material plane – for example, change its electromagnetic field as an active subject. To this the Earth said....

Yes, I can actively change my electromagnetic field.

But this is nothing that should be forced; it is a process of "approaching."

If an approach occurs too quickly it leads to fear and apprehension, even to enmity.

If it is just right it is an exciting unfolding.

Although the technical principle involved in the design of the device described by the Earth is fairly straightforward and simple, the technical implementation is not necessarily easy. Nor is it easy to hold the unwavering belief that contact will be achieved. In Chapter 9 I describe my own initial attempts at building such a device back in 1996/7 and the tentative results I achieved. In my opinion and experience, success in creating such a device is predicated on belief and desire. Personally, I no longer doubt both the possibility and feasibility of creating such a device, and I will continue my work to realize it.

Chapter 8

Awakening Unconscious Parts of You through Direct Interaction and "Lovemaking"

How Humans Use Their Energies

How do You humans use Your energies?

The correct or natural way
could be called "uplifting."

Naturally, the human being
lifts his/her energies upward;
but there is a blockage,
like a cloud cover hanging over him/her.

So therefore, since he cannot go up,
he also cannot go down to the roots.

Instead he goes outward, horizontally,
without roots and without source,
without the vertical axis, instead
"muddying the waters."

The axis is the key.

Human beings are social beings. Our main focus lies on our interactions with each other and on our material well-being. We focus on our relationships and our love lives, our jobs and careers, finances, health, politics, etc. This is what is here meant by focusing on the horizontal.

We are also universal beings, yet we seldom back away and look at the larger picture. We usually do not relate to the Earth under our feet, to the Moon and the Sun, to the vast Universe that surrounds us. When we do focus our attention on them, we do so mostly from a scientific, analytical point of view. We study the Earth in detail in order to determine how it can be used more effectively and to predict its processes, and we study cosmic processes as if they were void of consciousness.

The Earth is saying that we are living in the presence of alive and conscious processes, and if we connect consciously with the Earth and with the larger Universe surrounding us, we begin to expand and encompass more of who we are. This also leads to a different way of relating to each other, for we then relate to more of each other – we see each other in a deeper and more connected way, paving the way for more encompassing interactions.

The goal does lie in the horizontal;
but, by being fixated on it, humans behave
like a fly that tries to go through a window

by walking around on it for hours,
as if saying: "But I'm only 1/4 of an inch
away from the outside," instead of backing off
to see that the window is standing ajar.

The reason for this is shame.

Shame is primary.

Shame creates pain.

When the Earth states that the goal for us lies in the horizontal, this is similar to the statement, "Think globally, act locally." If we establish a "vertical" connection – i.e. if we open up to being consciously and intimately connected with the Earth, the Moon, the Sun, and the entire living Universe – our frame of reference changes, allowing deeper and more encompassing thoughts to guide our everyday lives.

Non-Identification

Whenever You are lonely, for example,
You are getting two messages.

One is that You are disconnected, and one is
that You are pulled toward connecting.

These two pull in different directions,
and they both seem to be true;
and yet they contradict each other.

When You feel disconnected,
You experience a kind of pain,
which is amplified through the realization of it.

This pain stops the reconnection.

The pain could also be called
identifying with the disconnectedness,
but there is no reason to feel pain
because You are disconnected.

Here is a crucial point
in all manner of change,
and it is a point where a new flow pattern
needs to be identified and created.

The feeling of disconnectedness
needs to bypass producing pain and go straight
to reveal itself as a pull to connectedness.

It is Your interpretation of Your inner habits
as "bad" when they are not wished by You
that causes problems.

The way out is not to break the habits,
but to look at them dispassionately
– with passion for something more immediate.
Love, for example, can give You
the right distance to anything else.

To some, this passage is self-explanatory; yet I would like to add my own take on it.

This has to do with the pervasive non-productive use of self-reflection. If we are moved to go in a certain direction – let us say toward more company and friendship, more money or more exercise – there is a tendency to redirect the energy of this impulse so it appears as a reflection of our own inadequacy.

The moment we translate a need, a want or a desire into a definition of ourselves as needy or wanting or lacking in some way – although this may seem to be technically true – we have first of all stopped focusing on what it is that we want. Instead, we focus our attention on ourselves, objectifying and identifying ourselves as a person with a problem. But the worst is yet to come. We then add to that an emotional judgment of ourselves as not being okay, not being whole. All we wanted was the healthy feeling of a wish fulfilled, yet through our way of dealing with this wish we have added a great impediment to its fulfillment. At the same time – and this is usually much worse – we have eroded our feeling of self-worth.

This is why it is easier to be a therapist or counselor to others than to counsel ourselves. It is not because the counselor sees better how to achieve what we desire, but because he does not see our wishes as a reason to judge us for not having attained them yet.

As I was working with this text, I realized that I, myself, had not truly taken this passage to heart. I had not applied it to my own way of dealing with my "issues" and "problems." It took quite some effort to begin to dispassionately develop the internal awareness necessary to see how I habitually translate many impulses or feelings into problems or "issues."

Lovemaking

As I was finalizing this book, I realized that the original title given for this chapter included the word "lovemaking," but the communications did not contain much about this topic. I asked about this and received the following answer....

Lovemaking is the purest form of co-evolution.

Not only can it result in procreation, which is
connected to Your evolution as a physical
species, but it also rejuvenates You
by clearing out old patterns
that have become stagnant and aligns You
with the deepest processes of evolution.

When You allow the energetic process of
lovemaking to encompass all of your energies,
from the raw power of sexuality to the
intimate flow of heart-to-heart communication
and the opening up of the mind
to the river of inner knowledge,
Your bodies begin to radiate from within.

* * *

This concludes my communications with the Earth, except for final words by the Earth in Chapter 11.

PART III

It Is Up to Us

Chapter 9

Subsequent Developments

Initial Experiments with the Communication Device

When the original communications with the Earth – which took place between September and November, 1996 – came to an end, I did not feel ready to publicize them. As I mentioned in Chapter 2, although I had promised the Earth to do so, I did not feel comfortable about going out with the information I had received, in part because I did not fully understand all of it and in part because I felt that it would not be taken seriously.

Instead, I decided to try to build the communication device described in the communications. I was excited by the idea of getting unequivocal messages from the Earth untainted by any mental and emotional distortions I might introduce. I also felt that if I could get such a device to work, any doubts I myself had about the validity of my communications would be removed. At the same time, I realized that this logic put the cart before the horse, since all my experiences told me that I would only succeed if I could first lay my doubts aside, and then build and test the device based on my deepest belief that it is possible and my deepest desire to succeed.

A friend of mine, a highly gifted electronics engineer, agreed to help me design the device, and we began by following the instructions in the communications. We took two copper rods, each two feet long, and placed them vertically into the ground fifteen feet apart. The bottom foot of each rod was in the ground and the top foot protruded into the air. We then measured the strength and fluctuations of the voltage between the rods with an oscilloscope. The difficulties came when dealing with the issue of translating the signals we received into 0s and 1s.

After trying out a few different setups, my friend developed the idea of using a carrier wave as an input into the rods, just as you have in radio and television transmissions, and placing a spiral antenna on the ground between the rods that would receive the signal from the Earth's electric field, including any possible messages, overlaid on the carrier wave. Later, we would filter away the carrier wave frequency from the amplified signal. Strictly speaking this was a deviation from the instructions received from the Earth, but we felt it might yield further insights.

What happened, however, was unexpected. The device was designed so that I could vary the carrier wave frequency (from about 15 to 30,000 Hz) and the amplitude of the input into the rods. The signal that the spiral antenna picked up was amplified and made audible through a loudspeaker. Since the carrier wave was much stronger than any signals in the Earth, what could be heard from the loudspeaker was more or less simply the amplified carrier signal. This sounded like a tone that I could change by varying the frequency of the carrier wave input into the rods. However, at certain frequencies and amplitudes, the signal received by the spiral antenna was not the constant tone of the carrier signal; instead it was an on/off signal. It very much resembled Morse code signals. I recorded these sounds on a cassette tape.

Since I had learned to send and receive messages in Morse code in the Swedish army, I tried to decode the signal as if it were a message in Morse code. However, in Morse code there is a slight pause between letters, and an even longer pause between words. The signal I received did not contain any such clear pauses. I put the project aside and since then have made no further attempts to decipher it.

I realized we had not yet built a device entirely according to the instructions, but at the time I simply did not have the resources to continue the project. I am now refurbishing the device to better conform with the instructions from the Earth and intend to continue this work in the near future.

My intent in building this communication device is not to prove that the Earth is alive. Even if such a device should begin to produce messages, it would be easy to come up with any number of scenarios whereby I could have faked these messages. Furthermore, I do not believe in the idea of trying to "prove" that the Earth is alive. Instead, as I mentioned, I believe the driving impulse behind building such a device must be that of desiring to communicate with the Earth. In addition, it requires the unwavering belief that such a device can truly produce messages from the Earth.

The Title of this Book

After 1996 I continued to try to "speak" to the Earth at times. Only sometimes did I feel that the connection was as strong as during the original communications. One time this happened was when I was working on a project to provide worldwide distribution of live images of the entire sunlit part of the Earth taken from a satellite. I was wondering if it would be good to have a title or a caption below the images, the way you might put a title under a piece of art. I asked the Earth about this, although I did not expect it to want any caption; but the answer came back quickly, "Yes, 'I Am With You.'"

At first I thought the title was a bit cheesy; but the more I pondered over it, the more I felt that "I Am With You" is a message that reverberates throughout the Universe – a statement of connection, presence and support that bridges worlds, between people, between all forms of life, between Great Spirit or God and humans, between planets, stars, and galaxies. It is a statement that connects, unites and brings together what belongs together. That is why I chose it as the title for this book.

How Was I Personally Affected by These Communications?

As I was writing this book, I felt as if I were coming home. I often connected deeply with the Earth and asked the Earth to help me write the book. It told me it was with me all the time, I should take my time, and I could let go of any feelings of loneliness or doubt and accept that I am held and loved by the Earth.

As I mentioned earlier, I received statements such as, "You are inside my heart." I slowly began to let this thought seep into my being. The image of being inside the heart of the Earth, which began as a thought, more and more became a feeling; and this feeling began to melt away a deep sense of estrangement – a feeling that we all, as human beings, are ultimately alone in a strange and foreign Universe. I realized that I hardly ever allowed this stark feeling of disconnectedness into my consciousness because it felt so terminal, so much a cold and terrible part of the human condition. When I finally began to let go of that feeling, it was replaced by a sense of gratitude at being at home on this warm, living Earth with whom we are all so deeply connected.

I also felt a growing excitement about the amazing, life-changing possibilities that open up once we together establish a reliable channel of communication, both internally and externally, with the Earth.

Chapter 10

A Sentient, Conscious, Communicating Earth?

The idea of a sentient Earth is incompatible with the currently held dominant worldview and scientific theory. But new scientific discoveries, insights and experiences are constantly adding to and changing what we regard to be true.

History is full of major changes in how we see ourselves and the world around us. One such change occurred when we established that the Earth is round and not flat. Another shift occurred when it was discovered that our bodies are full of bacteria and viruses which we cannot see with our naked eyes. The discovery of atoms and molecules, DNA and RNA are further examples. As we measured distances in space, realizing our solar system is part of our galaxy and there are billions of other galaxies in the Universe, our view of where we are in the larger scope of things was radically expanded.

But if the Earth is a conscious, communicating being, this will have even deeper consequences for our lives on Earth, for the statements by the Earth, taken at face value, go even further. If true, they represent a change in the meaning of our fundamental categories – such as time, consciousness and communication – and of the logic linking them together. This would then constitute a true paradigm shift.

But let us look at the normal stumbling blocks we have when contemplating the possibility of the Earth being alive and sentient.

Little attention has been paid to exploring the sentient nature of the world that surrounds us. The word "sentience" describes the ability to have sensations or experiences. This assumes the existence of consciousness, for without consciousness there is "nobody" there to have the experience. We humans have not yet developed undisputed methods to measure or "prove" the existence of consciousness. This is amazing, considering that consciousness lies at the heart of who we define ourselves to be.

The mere presence of certain electrical processes occurring in the brain does not define consciousness. On the other hand, we know that everything our brains do elicits changes in the electromagnetic fields of the brain. We also live on an Earth that is extremely active in terms of its electromagnetic field, which in turn is intimately connected to the electromagnetic activity of the Sun. Our whole bodies are suffused in this constantly changing electromagnetic field.

Consciousness research is increasingly moving away from seeing the origin of consciousness solely in terms of classical physics, such as the chemical and electrical processes occurring in the brain. Instead, the focus is shifting toward quantum mechanical processes. This would place the physical location of consciousness at the sub-atomic scale, which exists in all matter.

Most of us have never really considered the possibility that we are living in a world that is connected with other physical beings and things not only through direct physical interactions, but also from within. Every unborn baby is intimately connected with its mother's thoughts and emotions, a connection that remains once the child is born. Yet we quickly learn to accept that we are isolated beings, and what goes on in your head stays in your head.

New findings relating to how plants and animals communicate with their surroundings, including human beings, and studies of the processes occurring in the brain that reveal what occurs in the human subconscious, opens us up to new ways of understanding that we live in a sea of communication between all living things.

At the crossroads of life sciences, quantum physics and consciousness research – taking into consideration the experiences of those who have had studied the structure of consciousness from within – we are finding intriguing evidence of our abilities to communicate with other sentient beings from within.

The definition of life itself is also undergoing great changes. In its exploration of the Earth and beyond, science has been expanding its definition of life, contemplating the possibility of the existence of life with alternative biochemistries to that of life on Earth.

We are also but a small step away from considering the possibility that consciousness is not limited to the forms of life we know of so far. This is where the world of shamans and mystics has the potential to touch the mainstream of Western thinking, allowing us to open up to the greater realities surrounding us.

The idea that the Earth is somehow "alive" has been gaining influence. The "Gaia Theory" developed by James Lovelock[22] and Lynn Margulis[23], and the work of the deep ecology movement, combined with statements by representatives of various indigenous religions and spiritual paths, have brought this idea into the mainstream. But for the Earth to be seen as truly conscious – as a partner with whom a meaningful conversation can be held, not just for an individual but for humanity as a whole – a major shift needs to occur.

I believe the time has come for the idea that the Earth is a sentient and communicating being and we can tune in to and communicate with the Earth from within, both energetically and through conscious verbal communication. I also believe the time is not far off when we can speak to the Earth through a "phone" or a "video link," in English, or Japanese, or Swedish or Swahili. This will not happen suddenly; rather it will be a gradual, but not necessarily slow process.

It Is Up to Us

As I began to write this book and connected with the Earth, I received the following: "You are the determiner of the contact between us." I took this to mean it applied to me personally, and I needed to actively seek this contact for it to occur. Later, I asked for clarity regarding this statement and was told this indeed applies

---

[22] See *Gaia: A New Look at Life on Earth*, by James Lovelock. Oxford University Press, 2001.

[23] For example, *Symbiotic Planet: A New Look at Evolution*, by Lynn Margolis, Basic Books, 1999.

to me personally, but also to all individuals and humanity as a whole.

The Earth is thus not forcing its presence upon us, but is offering this communication as a possibility. Any communication requires the willingness of the participants to enter into communication. It is therefore up to us, individually and collectively, to allow for the possibility that the Earth is alive and conscious and is seeking to communicate with us consciously – not as a matter of faith, but as a possibility that we are willing to take seriously and pursue. If we do not pursue it, it will not occur. A phone that is ringing has to be picked up for a connection to be made.

This means that if conscious communication with the Earth is ever to become reality for more than a relatively small number of people, we must give the idea that such communication is possible the full benefit of the doubt. We must be fully open to the possibility, otherwise it cannot occur.

But we must do more than that. What is needed is a more active attitude than simply the temporary suspension of disbelief. It requires more than experimentation to prove or disprove the possibility; for it will only be activated if we activate it, not if we try to prove it. When we try to prove or disprove something, we are taking ourselves out of the equation in the name of objectivity. But establishing a line of communication with the Earth is a creative act on our part.

We must thus address the Earth with the intent of making contact; we must ask, even insist, that this contact be established. This is one of the things I learned during my extensive spiritual-shamanic training mentioned in my biography. Within many such traditions, the practitioner calls on various powers of the Universe to be present in order to create a container for whatever ceremony is to occur. Calling in these powers is of itself an act of power, an act that attracts them or manifests them; but it must be done with unwavering intent. The same seems to be true for communicating with the Earth.

This also applies to the building of a device whereby the Earth can communicate with us in words and images, and making it work is not just a question of getting it right technically. It must be based on the belief that it is possible, and it must be combined with a firmly held intent, a deep desire to establish this kind of contact.

The Irish-American researcher and philosopher Terrence McKenna once said, "The planet has a kind of intelligence; it can actually open a channel of communication with an individual human being."[24] I believe the day is not far off when such a channel of communication opens up for all human beings. But it is up to us to make it happen.

---

[24] Text by Terrence McKenna from the song *Re-evolution* on the album *Boss Drum*, by The Shamen, 1992.

Chapter 11

Final Words by the Earth

As I was finishing writing this book, I connected with the Earth and received the following addendum, or postscript, addressed to the reader of this book, spoken by the Earth....

I do not wish to hold monologues,
for a monologue is not true communication.

But it can open up the channel
for communication.

My "monologue" in this book
is more like the ring signal of a telephone,
in the hope that You "pick up the phone."

The bravery required on Your part – and it is
bravery – is to allow for the possibility
that we are deeply connected from within
and that a conscious connection,

both collectively and individually, is not only possible but also deeply desired on my part.

So If I have one single desire or request of the reader, it is that You go out into nature – or even a park – find a quiet place, undisturbed, and commune with me in whatever way comes naturally to you.

To paraphrase one of Your Hollywood movie endings, "This can be the beginning of a wonderful friendship."

Thank You.

You are invited to visit

www.IAmWithYou.com

for more information,

to leave a comment

or write a book review.

Lightning Source UK Ltd.
Milton Keynes UK
UKOW050854250212

187904UK00001B/11/P

9 783981 488913